Development of
AIR QUALITY STANDARDS

Edited by
Arthur Atkisson
Institute of Urban Health
School of Public Health
University of Texas

and

Richard S. Gaines
Director Air Conservation and
Environmental Health Programs
Southern Counties Planning Council

Environmental Resources, Inc.
Riverside, California

Charles E. Merrill Publishing Company
A Bell & Howell Company
Columbus, Ohio

International Standard Book Number: 0-675-09277-9

1 2 3 4 5 6 7 8 9 10 11 12 13 14 15-76 75 74 73 72 71 70

Printed in the United States of America

SYMPOSIUM ON THE DEVELOPMENT
OF
AIR QUALITY STANDARDS

A symposium Held Under the Auspices of:

National Air Conservation Commission
The Air Pollution Control Institute University of Southern California
The Christmas Seal Agencies of Southern California
Long Beach Tuberculosis and Health Association, Inc.
Tuberculosis and Respiratory Disease Association of Los Angeles County
Tuberculosis and Respiratory Disease Association of Orange County
Pasadena Tuberculosis Association
Tuberculosis and Respiratory Diseases Association of Riverside County
Tuberculosis and Health Association of San Bernardino,
Inyo and Mono Counties
Tuberculosis and Health Association of San Diego and Imperial Counties
Tuberculosis and Respiratory Disease Association of Ventura County
Tuberculosis and Respiratory Disease Association of California

October 23 to 25, 1969
Santa Barbara, California

DEVELOPMENT OF AIR QUALITY STANDARDS

INTRODUCTION

In 1968, following the designation of the South Coast Basin Air Quality Control Region by the California Air Resources Board, the eight Tuberculosis and Respiratory Disease Associations in Southern California formed a basin-wide Air Conservation and Environmental Health Committee. One of the first actions was the appointment of a Subcommittee on Air Quality and Surveillance to consider the role of the Associations in the setting of ambient air quality standards for California.

With the issuance of the first air quality criteria documents (particulates and sulfur oxides) by HEW, the subcommittee realized that there was very little time left for discussion. Under the terms of the Federal Air Quality Act of 1967, the states would be setting standards within six months from the date the criteria were issued. It was the consensus that a truly rational basis for setting standards had yet to be developed. For most states, health effects balanced against economic interests would serve as the basis for selecting specific numerical values for standards. Such a standard setting process ignores divers factors which should serve as input into any rational system.

Factors which must be considered include aesthetics, visibility, meteorology, behavioral responses, legal and social constraints, as well as pathophysiologic responses. For example, the ability to see the San Gabriel Mountains from the San Bernardino Freeway may be the most important air quality standard from the viewpoint of the Los Angeles residents. Despite provisions for public hearings by states which are setting standards, there is no adequate means to obtain input from citizens concerned with air pollution. The views expressed by those who participate in these hearings are often no more rational than the input from the industries which pollute.

Furthermore, the current standard setting process occurs in a vacuum. Very little consideration is given to the problems of implementation. Standards are set for single pollutants without regard to synergism; sources of pollution are not identified nor are they weighted in terms of importance. If a serious health effect for a given contaminant cannot be scientifically demonstrated, then a very weak standard or no standard at all may be set.

With this background, the Air Quality and Surveillance Subcommittee decided to convene a symposium to bring together air pollution scientists, engineers, lawyers, physicians, psychologists, political scientists and other professionals with an interest in the problem. The basic objective of this symposium was the stimulation of ideas for the development of rational models for setting ambient air quality standards.

The unstated objective was the hope that the decision making process could be affected by the ideas generated from a dialogue of professionals. Whether or not this unstated objective can ever be fulfilled can only be answered by time and an evaluation of the decisions made by the states.

This book is the outgrowth of one particular effort to affect the decision making process; it is part of an on-going process aimed at achieving an environment which enhances life rather than denigrating it. Air conservation is, after all has been said, more than an issue of health or survival, it is an issue concerned with quality of our lives.

The Christmas Seal Associations of Southern California wish to express their appreciation to the chairman and members of the Symposium Planning Committee for their many hours of effort in making the Symposium successful.

Symposium Planning Committee:

Sheldon Friedlander, PhD., Chairman, Professor of Chemical Engineering & Environmental Health Engineering, California Institute of Technology, Pasadena, California

Richard B. Atherton, Air Pollution Control Engineer, Ventura County Health Department, Ventura, California

James W. Daily, Petroleum Engineer, Air Pollution Department, Standard Oil Company, El Segundo, California

Richard S. Gaines, Director, Air Conservation and Environmental Health Programs, Tuberculosis and Respiratory Disease Association of Los Angeles County, Los Angeles, California

Burton Klein, Ph.D., Professor of Economics, California Institute of Technology, Pasadena, California

Clayton G. Loosli, M.D., Hastings Professor of Medicine and Pathology, School of medicine, University of Southern California, Los Angeles, California

Robert Pogrund, Ph.D., Associate Professor, School of Public Health, Center for Health Sciences, University of California

Loren Roberts, Executive Director, Long Beach Tuberculosis and Health Association, Inc., Long Beach, California

Seymour Schwartz, Doctoral Fellow, National Air Pollution Control Administration, Department of Industrial and Systems Engineering, University of Southern California, Los Angeles, California.

AIR QUALITY STANDARDS

TABLE OF CONTENTS

EVOLUTION OF AIR QUALITY CRITERIA AND STANDARDS

By John R. Goldsmith, M.D.
State of California Department of Public Health
Bureau of Occupational Health and Environmental Epidemiology

Introduction

Air quality criteria and standards are like a map and road signs. They provide description and direction which reflect and interpret reality, but they are not reality itself. The reality of air pollution must be recognized, apart from the maps and road markers, air quality criteria and standards, which derive from the experience with reality. They should help those who are unfamiliar deal with the air and its management. Just as there are many types of maps, so there are many valid descriptions of air quality. Legislatures or the Congress tell us what to cover in our criteria or maps.

This report is prepared from the point of view of a public health officer with an epidemiologic background. Obviously, air quality may appear to be something else to someone concerned with engineering, economics, law, or law enforcement.

Types of Evolution

As there has been an evolution of maps from the highly decorative but inaccurate Ptolemaic maps to the present photogrammatically precise cartography, so there is an evolution in air quality criteria and standards. Evolution has occurred on two major axes. First, there is the experience with other types of environmental exposures and their regulation, and secondly, there is the phenomena of air pollution, its perception, the reaction of individuals, the reaction of social groups, implementation of social decisions and some necessary abstractions. These are outlined in Appendix A and Appendix B. Appendix A-I shows the various aspects of environmental quality which have been subject to the same procedures which will lead to a set of air quality criteria and standards. They provide a more or less historical sequence from water, food, housing, air quality of the occupational exposure, and in the last few decades, radiation exposures. Of recently recognized significance is the interaction, for example; of housing involving water supply and waste management facilities, of the interaction of smoking, occupational exposure and of community air pollution in the genesis of chronic pulmonary disease and lung cancer; of multiple routes of lead exposure; and of the choice of radiation emissions or atmospheric pollution interacting with thermal water pollution aspects of power plant siting.

In the righthand columns of the Figure A-I are those attributes, reactions and implications of environmental quality which should be taken into consideration before criteria and standards can be derived. Certainly the health effects of pollution are of primary concern. Drinking water quality standards, perhaps the oldest and most effective from the health point of view were largely keyed to the avoidance of death from dysentery, cholera and typhoid fever and they have been successful. Even before the relationship of typhoid bacilli and water quality were known, the odor and taste of water were the indices which were used to determine the quality of the water supply. In the modern era, non-health effects are coming to assume a somewhat greater significance than they have in the past though their relative importance to health effects remains in general, secondary.

For any environmental exposure and effect relationship, the state of knowledge is extremely important. For certain types of relationships, the state of knowledge is certain and universal.

Concerning certain quality statements of exposure and effect, they are applicable wherever the phenomenon and receptors occur, for example, extremes of heat and cold are intolerable; there are radiation doses which are invariably fatal, even though at the other end of the spectrum it is difficult to define a dose low enough to have no effect whatever. For other exposures, while there is no doubt about the effect at a given dose, the individual or conditions under which the effect occurs are rather restricted; that is they are selective. For example, individuals of certain ages are more vulnerable to infection and to physical exposure than at other ages, and the individuals with certain herediatary attributes respond unfavorably to certain classes of drugs and occupational exposure. Some relationships are known to occur but only under special conditions. Such is the case for relationships of particulate and sulfur oxide pollution and chronic respiratory disease. Finally, there are certain types of reactions which are suspected but for which the state of knowledge really does not permit a reasonable inference of causation; for example, early changes in pulmonary or central nervous system function that can be tested for as a result of nitrogen oxide or carbon monoxide exposure. Based on animal data, they are thought to be possible. Of course, for a given type of exposure there remains the possibility that there are long-term consequences of repeated exposure, when the isolated exposure produces, only suspected, selective or trivial reactions.

The relation of air quality criteria and standards to other environmental areas are possibly closest with occupational exposures in the sense of having similar sets of data to deal with. Also, a very large number of occupational health standards are based on respiratory exposure as are many of those of importance to air quality. In the two kinds of populations, sensitivity is such that a broader range of reaction in relation to concentration of pollutant is to be expected for community air pollution than for occupational air pollution exposure to the same substance. The relatively short or small portion of the

2

working week in which most occupational exposures occur in relation to a period very much longer for community air pollution also distinguishes these two types of exposures. Finally, there is a great difference in the temporal-spatial relationship of exposure-effect and control procedure when occupational and community air pollution exposure are compared. Nevertheless, on the evolutionary ladder, occupational exposure experience is the nearest of all environmental programs to air quality criteria and standards.

The logical structure (the second axis) within which air quality criteria and standards have evolved is shown in the six charts of Appendix B. First, we must recognize that air quality criteria and standards are one of the forms of response to the perception of the air pollution phenomena. Phenomena themselves first have to be perceived and understood. Figure B-II. One must then understand the various ways in which pollution phenomenon can be responded to by personal reactions (B-III) which must be taken into account. Figure B-IV shows the social reactions which must also be taken into account. While it is often thought that air quality standards are to have their major impact in categorical-social restraints, such as by emission regulations or permits, it is also possible that emission charges, comprehensive quality regulation, taxation, and quality criteria planning and implementation with a cultural basis can be derived from this same standard-setting process.

The implementation of preventive and control procedures (B-V) does depend a great deal on certain abstractions, Figure B-VI, which are generally assumed and which have influenced the evolution of air quality criteria and standards. The understanding of these abstractions in different societies has given these criteria and standards a somewhat different character. For example, different societies have different views about property rights and about economic yardsticks as opposed to the other yardsticks. In some communities, a rational assumption concerning social processes is deeply imbeded in government and public behavior. In others, it is nearly absent. The more extensive analysis of such abstractions can document the extent to which there are interactions between different forms of environmental exposure and different forms of regulation for influencing them.

The experience of these two types of background can best be summarized by paraphrasing statements by Professor Abel Wolman. "Are the standards now being established with increasing numbers and perhaps decreasing validity, an adequate scientific and practical basis upon which the case for air pollution prevention and control can rest? Should they be met? Does the expenditure of air quality management dollars rest upon a reasonable balancing of convenience and of public service?" The original statement was made in 1940 with respect to sewage treatment, rather than air quality. Another commentary by Wolman which should provide a caution says, "The philosophy in most standards predominates that, in the absence of knowledge, the more rigid the standard, the safer the engineer". One must also accept the question raised by Wolman concerning the treatment of sewage effluents adequate to produce healthful environment for trout fishing in the following quote: "Given a limited tax dollar, does governmental discretion demand that it must be spent to permit a maximum of fish life in a receiving body of water at the sacrifice of houses, hospitals, food, water supply, smoke prevention, and so forth on land?" If the cost of preserving the amenities of the land, and the water environment, were so low that all could be accomplished with satisfaction, none of these issues need to be raised. This condition of adequate resources for all environmental quality purposes did not prevail, does not now exist, and is not likely to appear in the near future. The issues remain pertinent.

Thus we can infer from these experiences with other environmental pollution problems that a great deal of the uncertainty lies in whether the meeting of standards is justifiable in terms of total resources available and their allocations to a variety of purposes. However, and this must be strongly emphasized, the decision concerning the relative values of different programs does not lie with the scientist or the engineer, but does lie with the political decision-making and appropriating body. True, it is appropriate for the scientist and engineer to put forth as clearly as possible the knowledge concerning exposures and the effects. How these statements of criteria are to be reflected in community air quality goals is partially a political process. It is principally a political process which determines the urgency of competing goals for public resources and to some extent, for private resources as well.

The Evolution of Air Quality Criteria and Standards in California

For many years, the Division of Environmental Sanitation of the California State Health Department had been studying and promulgating environmental standards for food, water, and sewage treatment and participated in or observed the establishment of occupational and radiation health standards. Mr. Frank Stead, Chief of the Division, first formally suggested the desirability of air quality standards with multiple levels which reflected the diversity of air pollution effects.

A review of the background of other environmental health standards as was made in 1959 and published in the Technical Report on California standards for Ambient Air Quality and Motor Vehicle Exhaust, pages 18 through 23. This included certain basic principles applied in water sanitation and radiation standards, industrial exposure standards, a somewhat more detailed presentation of the Los Angeles Alert Levels, a discussion of community air pollution standards in the Soviet Union and the commentary on the few other occasions where air quality standards had been set at that time. The same report contains a chapter on the theory of standards for air quality, which is relevant to this symposium but need not be reviewed. (The second edition of Stern's "Air Pollution: A Comprehensive Treatise" (1968) contains 56 pages (!) of tabulated Air Quality Standards. Standards have proliferated like rabbits!)

Quotation from Wolman were from the collected papers, entitled Water, Health, and Society, pages 109 and 111. Indiana University Press, 1969. (G.F. White, Editor)

The Los Angeles Alert System has often been confused with air quality standards. The alert levels had as their purpose the agreement on a quantitative yardstick by which local and state authorities would be committed to take certain actions to intervene in the event that air pollution levels appear to be increasing up to a point which might have important and immediate consequences on health. It follows from this that the alert levels took no cognizance of possible long-term effects. The State Health Department thought that this was a reasonable plan for coping with possible air pollution emergencies and reviewed the proposals in 1955 and in conjunction with the U.S. Public Health Service, Committee of Toxicologists in 1958. While the scientific committee of the Los Angeles County Air Pollution Control district has recommended some changes in the alert stages, particularly for carbon monoxide, the alert levels have not been changed.

A similar yardstick was less formally adopted by a similar mechanism in 1969. On a recommendation of the Los Angeles County Medical Society, the District and the public school system in Los Angeles County agreed that when oxidant reached a concentration of 0.35 ppm, they would recommend that the school children not be allowed to engage in strenous physical activity because of possible risk to health. This recommendation is more in the nature of an alert level than it is in an air quality standard or an air quality criteria, since no specific evidence is available to direct attention to this particular number and the procedures which were followed were not those used for air quality criteria and standards.

Impetus was given to the development of air quality standards through the recognition in 1958 that photochemical air pollution problems could not be solved without control of motor vehicle emissions. While the Los Angeles district is admirably suited for the control of fixed sources, it was not so well-suited for the control of mobile sources. It was becoming equally obvious in 1958 that motor vehicular pollution was a serious problem in other parts of the State. Governor Brown in his special message to the Legislature on February 10, 1959 said "The time has now come for California to take the lead and establish standards for the purity of the air. Standards for safe air will give local control officials and health officers a measuring stick for smog. It is essential that we know what level of pollution threatens death or illness, or impairs the health of our people. Unless we establish these guideposts, we risk our happiness, our health, and indeed our lives."

In response to this request, the 1959 Legislature enacted the following addition to the Health and Safety Code, Section 426.1:
"The State Department of Public Health shall, before February 1, 1960, develop and publish standards for the quality of the air of this State. The standards shall be so developed as to reflect the relationship between the intensity and composition of air pollution and the health, illness, including irritation to the senses, and death of human beings, as well as damage to vegetation and interference with visibility."

5

(The statement concerning the relationship between time and concentration and specified effects are now based on "Technical Reports on Air Quality Criteria" and are called Air Quality Criteria.)

Scientific Basis for Standards

A substantial body of information, derived for the most part from industrial toxicology, existed in 1959 concerning the health effects of single substances. In-plant contaminant concentrations were usually expressed in threshold limit values for eight-hour exposures of healthy workment and the levels chosen were based on a number of different cirteria; for example, a level which produces only minor acute health effects or injures a very small proportion of workmen, or a level which is believed to cause no illness or toxic effect during a lifetime, or a level which need not be exceeded because of technologic reasons, or a level which is not uncomfortable to high proportions, say 90 percent, or previously unexposed workmen.

Policy Decisions of the California Health Department

Associated with this body of information was a change in emphasis from concern about the production of unique or specific disease to concern about air polution as a factor which might aggravate pre-existing disease or contribute to the development of diseases which might also be caused by non-air pollution exposures and conditions. There was a shift of concern from the effect of community air pollution as a specific toxicant, to its effect as an agent which may impair health. Impairment of health was interpreted to include irritation to the senses.

In response to the Legislature's directive to adopt air quality standards, the Departmental staff developed a series of proposals. These were then submitted to the Department's Advisory Committee on Air Sanitation and the Subcommittee on Ambient Air Quality Standards and Research. Five of these proposals were:

1. The Department should concern itself with all types of air pollution which could produce the effects stipulated in the Legislative directive. Its concern should not be confined exclusively to the photochemical pollutants.
2. The complete abolition of air pollution is impossible and the objectives of the standards are to set levels at which specified effects or their absence can be predicted.
3. Any standards set must be based on sound data and concurred in by scientists in air pollution and related fields.
4. The standards relating to human health and well-being should be based on the groups or persons in a population who are most sensitive to air pollution effects, provided such groups be definable in terms of age and medical status.

6

5. Since air pollution can produce a multiplicity of effects, the standards should be set at several different levels.

At that time, a three-level scheme was adopted.

Choice of Three Levels and Their Definitions

This list of possible and measurable air pollution effects was assembled:
1. Acute sickness or death.
2. Insidious or chronic disease.
3. Alteration of important physiological function.[1]
4. Untoward symptoms.[2]
5. Discomfort from air pollution sufficient to lead individuals to change residence or place of employment.
6. Damage to vegetation.
7. Loss of visibility.

It was decided to define three levels of air pollution standards.

1. *"Adverse"* Level –– The first effects of air pollutants are those likely to lead to untoward symptoms or discomfort. Though not known to be associated with the development of disease, even in sensitive groups, such effects are capable of disturbing the population stability of residential or work communities. The "adverse" level is one at which eye irritation occurs. Also in this category are levels of pollutants that lead to costly and undesirable effects other than those on humans. These include damage to vegetation, reduction in visibility, or property damage of sufficient magnitude to constitute a significant economic or social burden.

2. *"Serious"* Level –– level of pollutants, or possible combination of pollutants, likely to lead to insidious or chronic disease or to significant alteration of important physiological function in a sensitive group. Such an impairment of function implies a health risk for persons constituting such a sensitive group, but not necessarily for persons in good health.

3. *"Emergency"* Level –– level of pollutants, or combination of pollutants, and meteorological factors likely to lead to acute sickness or death for a sensitive group of people.

Organization of Data

When a substance, or group of substances, had been selected for consideration, the staff of the Department either prepared or arranged for the preparation of a report containing the relevant information from the literature on concentrations and effects of substances, the sources from which the pollutant comes, the time course of the pollutant, and the likelihood and

---

[1] By important physiological function was meant function such as ventilation of the lung, transport of oxygen by hemoglobin, or dark adaptation (the ability to adjust eye mechanisms for vision in partial darkness).

[2] By untoward symptoms was meant symptoms which in the absence of an obvious cause, such as air pollution, might lead a person to seek medical attention and relief.

characteristics of unusually sensitive groups of people or categories of plants. The crucial statements concern the relationship between time and concentration, and specified effects.

(These statements relating exposure and effect are now called "criteria".)

What the 1959 Standards Were Not

The standards set for single pollutants alone were not necessarily applicable to combination of pollutants, to physiological effects under unusual weather conditions, or to pollutants occurring in aerosol form or when in combination with aerosols. For these reasons, no single level of a given pollutant, ordinarily without effect at that level, was interpreted as a guarantee that levels below that point were safe or free from effect.

The history of public health research contains many examples in which materials originally thought to be innocuous were later shown to be of substantial hazard or harm. There was no reason to expect our knowledge of air pollutants to differ. For these reasons, the standards of air quality were not intended to provide a sharp dividing line between air of satisfactory quality and air of unsatisfactory quality, but only to indicate approximate levels at which certain definite effects could, on the basis of current information, be anticipated.

Planning Implications of Standards

The use of air quality standards in urban planning or zoning has been pointed out by Holland, et al., who presented an equation which signifies one importance of standards. The equation is as follows:

$$\begin{matrix} \text{Contaminant} \\ \text{Emissions to} \\ \text{be Permitted} \end{matrix} = \begin{matrix} \text{Dilution} \\ \text{Capacity of} \\ \text{Air Shed} \end{matrix} \times \begin{matrix} \text{Community Standard} \\ \text{for Desirable} \\ \text{Air Quality} \end{matrix}$$

This equation implies that with air quality standards, and a knowledge of the dilution capacity of the atmosphere the quantity of wastes that may be discharged can be determined. One should not underestimate the difficulties of obtaining accurate estimates of atmospheric dilution capacity, allocation of given volumes of air among polluting sources, and even of the metamorphosis of pollutants which may occur in the atmosphere.

A criterion which needs to be considered for inclusion in the basic scheme of air quality standards is the level below which no biologically significant or esthetic effects can be detected. Such "no effect" levels are the bases for air quality standards in the Soviet Union. This criterion was not used in California because it would seem to require control that might be impossible or unreasonable to achieve. Also, it tends to imply that compliance with such a standard is necessary to avoid disease.

A range of standards (criteria) representing a series of predictions as to consequences not necessarily associated with morbidity and mortality in the classical sense is a new idea in public health. Nevertheless, it was just such an extension of these standards that was recommended by the World Health Organization's Inter-Regional Symposium on Air Quality Criteria in August of 1963.

A definition of these criteria may be found in the Report of the World Health Organization Expert Committee on Atmospheric Pollution (WHO Technical Report #271) as follows: "Criteria for guidance to air quality are the tests which permit the determination of the nature and magnitude of the effects of air pollution on man and his environment." "Guides to air quality are sets of concentrations and exposure times that are associated with specific effects of varying degrees of air pollution on man, animals, vegetation and on the environment in general". The committee then went on to recommend four categories of concentrations, exposures, times, and corresponding effects which were the three levels used in the California Air Quality Standards, plus a lower level which was defined as "concentration and exposure time at or below which according to present knowledge, neither direct nor indirect effects (such as alteration of reflexes or adaptive reactions) have been observed".

It is clear that in light of the WHO definition what the California State Health Department did in 1959 was to call "air quality standards" something which the World Health Organization calls "criteria and guides to air quality".

Subsequently, the Public Health Service has used essentially the same definition under the term "air quality criteria". These are defined in an enabling act by the Congress:

"Whenever the Secretary of the Department of Health, Education, and Welfare determines that there is a particular air pollution agent (or combination of agents) present in certain quantities, producing effects harmful to the health or welfare of persons, the Secretary shall compile and publish criteria reflecting accurately the latest scientific knowledge useful in indicating the kind and extent of such effects which may be expected in the presence of such air pollution agent (or combination of agents) in the air in varying quantities".

And further in Public Law 90-148, "Such criteria shall accurately reflect the latest scientific knowledge useful in indicating the kind and extent of all identifiable effects on health and welfare which may be expected in the presence of an air pollution agent, or combination of agents, in the ambient air, in varying quantities".

The Mulford-Carroll Act of the California Legislature, establishes an Air Resources Board, and (Section 39051, Part 1, Division 26 of the Health and Safety Code) directs the Board among other things to "Adopt standards of ambient air quality for each basin in consideration of the public health, safety and welfare, including but not limited to health, illness, irritation to the senses, aesthetic value, interference with visibility, and effects on the economy. These standards may vary from one basin to another. Standards relating to health effects shall be based upon the recommendations of the State Department of Public Health".

Note that aesthetic value and effects on the economy are added to the "criteria" to be considered in setting the standards. Also novel was the permission to have varying standards in different basins. The Board is also directed to fix the boundaries of the basins by January 1, 1969.

Results of the Evolution

Thus the Department of the Health, Education, and Welfare is the authorized agency of the Federal Government for issuing Air Quality Criteria. Its position and its scientific basis is presented by the reports of the National Air Pollution Control Administration. These were further discussed in the Congressional Record of hearings before the Subcommittee on Air and Water Pollution of the Committee on Public Works, U.S. Senate, 90th Congress, July 29, 30, and 31, 1968.

The relationship of scientific capabilities and judgment to the Federal task is still not entirely clear. Its criteria for sulfur oxides and for particulate matter have been published in February, 1969.

The States in which Air Quality Control Regions have been designated, must within a certain number of days of the issuance of criteria, adopt air quality standards, and then implementation plans.

The table of standards now being adopted in California is shown as Table I and is based on the technical documentation prepared by the Air Resources Board and by the Department.

The World Health Organization continues to be interested in international agreements on air quality criteria and guides. Its interest tends to prevent any one set of assumptions to exclusively dominate the work on this problem in the U.S. In the WHO forum, there will continue to be raised the question of a "hygeinic" level of air quality or a no-effect level, even though much of the work in the U.S. on standards will involve setting achievable goals. This may lead to our use of the term "air quality standards" in a different sense than the phrase is used in other countries.

Thus there is every expectation that concepts, facts, and their application will continue to evolve.

Air Quality Standards – Future Evolution?

We have still a number of air quality standards that are as decorative as Ptolemaic maps, and just as hazardous to use as a guide to private and public voyaging into the realm of air conservation.

Air conservation should now be our goal. It has been systematically presented by the Air Conservation Commission of the AAAS. It is being sought principally by voluntary agencies, especially the Tuberculosis and Respiratory Disease Association which sponsors this meeting.

10

Table 1

Proposed And Present Ambient Air Quality Standards

Pollutant	New or Proposed Standard	Objectives of Standard	Former Standard
Oxidant	0.1 ppm for one hr.	to prevent eye irritation and possible impairment of lung function in persons with chronic pulmonary disease. Also to prevent damage to vegetation.	0.15 ppm for one hr.
Carbon Monoxide	20 ppm for 8 hrs. (50 ppm for 1 hr.)*	to prevent interference with oxygen transport by the blood based on carboxyhemoglobin levels greater than 2%.	30 ppm for 8 hrs., or 120 ppm for one hr.
Sulfur Dioxide	(0.1 ppm for 24 hrs., when particulate matter standard is exceeded	to prevent possible increase in chronic respiratory disease and damage to vegetation	1 ppm for one hr., or 0.3 ppm for 8 hrs.
	0.5 ppm for one hr.)*	to prevent possible alteration in lung function and irritating odor	—————
Particulate Matter	(100 $\mu g/M^3$ arithmetic mean of representative 30 days of consecutive 24 hr. samples	to improve visibility and prevent acute illness when present with about 0.05 ppm sulfur dioxide	—————
	visibility of not less than 7.5 miles when relative humidity is less than 70%)*	to improve visibility	visibility of not less than 3 miles when relative humidity is less than 70%).
Hydrogen Sulfide	0.03 ppm for one hr.	to prevent offensive odor	0.1 ppm for one hr.
Nitrogen Dioxide	0.25 ppm for one hr.	to prevent possible risk to public health and atmospheric discoloration	0.25 ppm for one hr.

*Not yet adopted by Air Resources Board. Others adopted 17 September 1969.

On November 19, the Air Resources Board adopted a 24 hour SO_2 standard of 0.04 ppm, a 1 hour standard of 0.5 ppm of suspended particulate matter standard of 60 $\mu g/M^3$ as an annual average, one of 100 $\mu g/M^3$ as a 24 hour average and a visibiltiy standard of 10 miles.

11

The future must also include better surveillance of air quality deterioration and related effects, better understanding of the factors of safety with which criteria and standards can be related, and a better focus on and support of research on air pollution effects in light of all the factors which interact to influence them.

EVOLUTION OF AIR QUALITY CRITERIA AND STANDARDS

Appendix A

Appendix A-I

Bases For Environmental Criteria And Standards

Environmental Vector	Vector Control Classification	Factual Background For Control Decisions	Relevant Classifications
Water	Drinking Bathing Sewage Rx.	Health Effects	Mortality Morbidity Irritation Impairment
Food	Milk, etc. Quality Preservation Additives	Non-health Effects	Visibility Economic Loss Ecological change
Housing	Safety Hygiene Noise Crowding	State of Knowledge	Certain and universal Certain — selective Possible—Univers Possible—Select Suspected
Occupation	Safety Irritants Toxicants	Control Strategy	Process change Emission control Dilution Removal Specific pro- tection
Radiation	Community Occupation Therapeutic	Temporal-spatial Impact of Control	Immediate Temporary Delayed Decremental

Appendix A-II

Bases For Air Quality Criteria And Standards

Environmental Vector	Vector Control Classification		Relevant Classifications
Air	Gases		Environmental phenomena
			Non-health effects
	Aerosols	Criteria	
			Experimental (toxicological)
	Indices and Attributes		Experiential (epidemiological)
			Certainty, severity and Universality of Criteria
	Source group	Standards	Ease and cost of control and secondary factors
			Temporal-Spatial considerations
			Value judgements. (safety factors)

EVOLUTION OF AIR QUALITY CRITERIA AND STANDARDS

I.   Phenomena — What is happening
     A.   Emission of pollutants as a function of time and location
     B.   Atmospheric attributes in relation to topography as a function of time
     C.   Alterations in pollutants in atmosphere: Transform of A by B
     D.   Direct effects of pollutants, modified by C on physical systems, chemical systems, biological systems — (vegetable, animal, man)
     E.   Indirect effects of pollutants modified by C, X (other phenomena), and displaced in time and space

II. Preception of Pollution Phenomena
     A.   Direct sensory perception, visual, auditory, olfactory, (thermal, tactile), as a function of acuity, time, and external and inherent background
     B.   Indirect perception — irritation, avoidance, adaptive reactions, conditioned reflexes
     C.   Mediated perception — by instruments, by direct perception of indicators, by inference, by research

III. Individual Reaction to Phenomena as Conditioned by Perception
     A.   Conscious or unconscious dilution or avoidance on an individual basis
     B.   Personal or family dislocation
     C.   Modification of emissions spontaneously by those responsible
     D.   Personal or family protection from exposure

IV. Social Reaction to Phenomena as Conditioned by Perception and Individual Reaction and Influenced by Communications.
     A.   Complaint or objection through social channels.
     B.   Categorical social restraint by permit, by emission regulation, by process regulation, by design specification.
     C.   Flexible social restraint by emission charges, by comprehensive quality regulation, by taxation
     D.   Cultural adaptation by establishment of quality criteria, planning and, implementation of quality-seeking behavior.
     E.   Organized effort to seek needed information.

V.   Implementation of Social Reactions
     A.   Of process control on basis of cost-free choices
     B.   Arriving at criteria
     C.   Relating standards to criteria in light of local factors
     D.   Allocation of cost of remaining control and other costs needed to achieve C

E.   Achieving standards (implementation plan) through process alteration, control, dispersion, or dislocation

F.   Feedback of information and alteration of implementation plan

VI. Abstractions Implicit for Operational Success of Implementation.

    A.   Rights and obligations of individuals, technical process operators, governments at various levels, non-governmental organizations, scientists, research and educational institutions.

    B.   Rationality assumption concerning effectiveness of social processes with respect to behavior in light of present and anticipated costs and benefits.

    C.   Scientifically and rationally consistent system of statements concerning predicted effects of pollutants ("criteria") and desired upper limits of quality impairment ("standards").

    D.   Responsiveness of social system and statements (VI-C) to new information concerning attributes and reactions of entire system.

Discussion by D.S. Barth, Ph.D., of the Paper
"Evolution of Air Quality Criteria and Standards"

by John R. Goldsmith, M.D.

The analogy of "air quality criteria and standards" to "a map and road signs" is not a particularly good one. The term "criteria" is defined on page 14 of the report by a quote from Public Law 90-148, to wit, "Such criteria shall accurately reflect the latest scientific knowledge useful in indicating the kind and extent of all indentifable effects on health and welfare which may be expected in the presence of an air pollution agent, or combination of agents, in the ambient air, in varying quantities". Unfortunately the term "standards" is not equally well defined.

Whenever a discussion of standards with regard to air pollutants is being undertaken it is essential at the outset to differentiate between ambient air quality standards and emission standards. In general ambient air quality standards are concentrations of specific pollutants, in appropriate mass per unit volume units, which may not be exceeded at any point within a defined geographical region more than a specified number of times in a given time period. Emission standards, on the other hand, are amounts of emitted specific pollutants, expressed in appropriate units, which may not be exceeded by any emission source. In addition the definition of any air pollution standard must include a definition of the measurement technique which will be utilized to determine compliance.

Apparently Dr. Goldsmith intended to discuss principally ambient air quality standards in his paper. His intent in this regard should have been clearly stated to avoid any possibility of ambiguity. Ambient air quality standards are essentially goals which are set on the basis of criteria. Except for the case of single isolated air pollution sources it is usually not possible to rigidly enforce such standards by law. Emission standards are enforceable by law and are the normal means utilized to bring about the achievement of desired ambient air quality standards.

In view of the foregoing discussion let us now examine the analogy of "air quality criteria and standards" to "a map and road signs". The word "standards" here apparently refers to ambient air quality standards. The description provided by the map may correspond to criteria and ambient air quality standards; however, the road signs and the roads themselves show how to proceed from one point to another. In this sense the roads are more nearly analogous to *emission standards* since they indicate how one should proceed to achieve a certain goal. Thus I submit that the roads correspond more to an implementation plan with its included emission standards than to ambient air quality standards.

The statement, "The relationship of scientific capabilities and judgment to the Federal task is still not entirely clear." which is found at the bottom of page 14 is difficult to interpret. We believe that the mechanisms set up for the development of Federal criteria and control techniques documents and the subsequent use of these documents by the States to set ambient air quality standards and implementation plans have been clearly defined and are understood. The clear intent of the Clean Air Act as amended is that the control of air pollution at its source is the responsibility of State and local governments. The Federal criteria documents provide the States with a rational scientific basis for the development of ambient air quality standards. Within certain defined limits then it is the task of the State and local governments to exercise judgment with regard to the setting of precise ambient air quality standards. This judgment must take into consideration such factors as public attitudes, economic feasibility,* existing background levels of the pollutants of concern, effects on future economic development, legislative authority, availability of resources to implement any plans, etc. The state and local governments are certainly in a better position to evaluate such factors than would be the Federal government.

The Section on "Air Quality Standards — Future Evolution" beginning on page 15 is quite short and unsatisfying. It leaves more questions in the mind of the reader than it provides answers. If one attempts to look ahead to the year 2000 A.D., what should be the answers to the following questions:

1. Will the present system of publishing criteria documents for specific single air pollutants continue to be adequate?

2. How many additional criteria documents will be required and for what specific substances?

3. In the development of standards what weighting factors should be used for the different kinds of effects described in the criteria?

4. How does one go about developing "better surveillance of air quality deterioration and related effects?" Specifically how does one determine the optimum number, type and location of sampling devices and the optimum sampling frequency; How should the resulting data be handled and displayed?

5. How can we achieve "better understanding of the factors of safety with which criteria and standards can be related"?

6. Specifically what "research on air pollution effects" needs to be done and in what priority? How should the available resources be distributed over research on human health and welfare, economic, aesthetic and geophysical or global effects?

Answers to such questions and many more seem to be fundamental to defining the possible future evolution of air quality criteria and standards.

---

*This factor should not be considered where protection of public health is at stake.

Discussion of Evolution of Air Quality Criteria and Standards

By James H. Sterner, M.D.
Associate Dean and Professor of Environmental Health
The University of Texas School of Public Health
P.O. Box 20186
Houston, Texas 77025

Dr. Goldsmith's eloquent presentation has traced the evolution of air quality criteria for the control of air pollution and noted that "occupational exposure experience is the nearest of all environmental programs to air quality criteria and standards." He implied, but did not expressly state that the problems associated with the development of these criteria for the control of community air pollution are inherently more complex and more difficult of solution than those encountered in the industrial health setting.

"Air quality criteria" for hazardous physical and chemical agents of industrial interest, have made possible the phenomenal technological expansion of our industry with a cost-benefit ratio which has been acceptable to society, and in relation to the potential hazard a remarkable achievement. Admittedly, the application in the protection of our labor force has been uneven, with the larger companies benefiting from their ability to establish better industrial hygiene and safety programs, and with effective governmental surveillance of working conditions limited only to a few states and cities with official occupational health units adequate for the job.

The earliest control of occupational exposures was by a trial and error process. If workmen developed an illness identified with a given set of working conditions, efforts were made to reduce the exposure, frequently by a stepwise procedure, until a set of conditions was achieved which did not result in obvious illness. In addition, many occupational diseases were characterized by distinctive signs and symptoms, making the clinical identification an easier task.

As analytical methods developed, it was possible to make reliable environmental measurements of the levels associated with injury, and as conditions were improved, to identify the levels which did not result in recognizable occupational disease. In many instances, it was technologically and economically feasible to still further reduce the exposure, and thus add a further "factor of safety" to the threshold limit value.

This movement downward, in the establishment of certain occupational air quality criteria, from levels associated with overt disease, thru levels accompanied by lesser symptoms, to levels of "no effect", permits a more orderly and probably more reliable determination of criteria. In the process of establishing criteria for the control of community air pollution presently encountered in our cities, we do not have the clear and convincing evidence found in many instances of occupational exposure. The lessons learned from the

21

"episodes" of London, Donora, New York, and Los Angeles are quite prophetic, but the evidence for clearly defined effects at the more prevailing conditions is much more difficult to interpret. But "difficult" or not, I firmly believe that the immediate establishment of air quality criteria is essential to a fair, equitable, and effective control program. An alternate course to guide and direct the control mission would require the conferring of an unprecedented degree of authority to our control agencies, with the possibility that arbitrary or even capricious actions might jeopardize or unduly delay through legal recourse the whole control program. It might be relatively easy to move against the more obvious offenders in a community, but the elimination of the overt and more easily identified polluters will not solve the ambient air problems of our larger communities.

The far-reaching effects of the air quality criteria, and their reflection, along with economic, political, sociological, technological, and ecological factors in the establishment of standards, requires our serious attention and consideration as to the mechanisms by which such important guides are and should be developed. The earlier effort by the Department of Health, Education and Welfare to develop sulfur oxides criteria as an essentially in-house activity resulted in such a violent reaction that Congress, in later legislation stipulated that a more broadly based body be developed, and the National Air Quality Criteria Advisory Committee was established. In turn, subcommittees were formed to deal with specific pollutants, and criteria have been issued for particulates and for sulfur oxides. We are now in the testing period for the interpretation of the "summary and conclusions" of these reports, as state by state, the criteria are transmuted into standards. There are honest and intelligent differences of opinion among knowledgeable people viewing the same data and extrapolating to standards. We are in an interesting time of resolution, with many uncertainties and many gaps in the web of evidence we would like to have to make such significant determinations. Unfortunately, time is running out for us to take action, for each year the atmosphere gets dirtier, and many thoughtful observers believe that we have passed or shall soon pass, the point where serious health effects will be only too evident.

With the issuance of the first two criteria, for particulates and for sulfur oxides, the interrelation of the pollutants has been emphasized. The problem becomes increasingly difficult as we attempt to establish values for other important elements of pollution. Here, as with carbon monoxide and lead, we encounter other sources of exposure than community air pollution, and other means of body intake than by inhalation. It is likely too that special industrial contributors such as smelters, or as in Houston, a concentration of the petrochemical industry, may significantly change the character of the biologic effect.

In establishing air quality standards, we shall certainly have to give constantly increasing consideration to the changing miliue for man, with a whole battery of increasing stresses, real and potential and largely unassessed. We know so little of the effects of minimal insults spread over large populations, of whether two or more such factors are additive, substractive, synergistic, or anergistic. Very few if any of these widely pervasive but potentially hazardous agents are pathognomonic, that is, producing disease which is sufficiently distinctive and characteristic that an effect can be ascribed to a particular agent. An increasing number of chemical substances are identified as having carcinogenic, mutagenic or teratogenic effects, admittedly usually at higher levels of administration. We must have concern about the hazard, but we should not deprive society of needed and desirable benefits on inadequate evidence or emotional caprice. More and more important judgments as to the hazard of agents will be required, and as we extend our considerations to longer and more difficultly determined end-points such as carcinogenisis, longevity, and genetic effects we shall have to make our judgments with less adequate data.

Some time ago simply as an example of the changing milieu I noted a yearly production figure of some 3 billion pressurized spray units, of which 2½ billion were for use in the United States. I emphasized the fact that I had no evidence of harm from the use of such materials, but with the increasing volume and variety of chemicals disseminated, there is reason for concern. If a substance such as beryllium, in a form which could produce berylliosis, were distributed thinly and widely thru the population, recognition of the hazard might be very long delayed. Again, I want to stress that I'm not against pressurized spray materials — I use them too — but an average consumption of ten cans per person per year cannot be ignored as a potential source of hazard.

There will be an increasingly greater need for the development of criteria relating biological effect to an ever greater variety and volume of potentially hazardous physical and chemical agents. Some of these criteria will be concerned with the general population, others with special populations such as industrial workers, but in most cases there will be an overlap of interest. The establishment of a more broadly based National Air Quality Criteria Advisory Committee for the development of criteria was an improvement over the initial effort to set such levels internally in the agency responsible for the control program. Since acceptance of the criteria will be of extreme importance in the effectiveness of any control program, it would seem desirable to have the body responsible for developing criteria, as broadly based, as objectively representative as possible, of the knowledgeable and expert individuals.

The success of the National Council on Radiation Protection and Measurements in the field of radiation criteria suggested a model for dealing with other hazardous physical and chemical agents. The NCRP is composed of representatives from those professional and scientific societies recognized for their interest and competence in matters of ionizing radiation effects and protection. A similar organization for other hazardous physical and chemical agents could develop needed criteria by establishing special or ad hoc committees composed of experts in the particular subject areas, with the main council serving a balancing and integrating function, increasingly necessary as effects from diverse hazards interrelate and overlap. Such a mechanism would insure the greater participation and involvement of the scientific and professional

23

community, and add recogniction and sanction to the findings and recommendations. The responsibility for transmuting the criteria into standards would still vest the governmental bodies.

The proposals for the development of such a National Council on Hazardous Physical and Chemical Agents has received enthusiastic endorsement from several professional and scientific organizations. Preliminary discussions have been held with the National Academy of Sciences — National Research Council to consider the possibility of that organization incorporating such a council, or serving as a base for such a function. While the present constructive action in the development of air quality criteria must maintain its momentum, there may be considerable merit in broadening the base of participation and acceptance through such a mechanism as suggested.

MODELS FOR AND CONSTRAINTS ON DECISION MAKING

By Seymour Schwartz
Doctoral Fellow, National Air Pollution Control Administration,
Department of Industrial and Systems Engineering
University of Southern California

Gilbert B. Siegel
Associate Professor of Public Administration
University of Southern California

ACKNOWLEDGMENT

We are indebted to Arthur A. Atkisson, Director, Institute of Urban Health, School of Public Health, University of Texas, for many valuable ideas, suggestions, and hours of stimulating discussion of topics related to this paper.

Mr. Schwartz is grateful to the National Air Pollution Control Administration for its generous support under fellowship 5 F3 AP 38,732-03.

I. Introduction

The deterioration in the quality of urban environments is rapidly making them unlivable. Stresses from pollution, noise, traffic congestion, crowding, and blight are reducing our capacity to enjoy life and threatening our mental and physical health. Progress in halting this disastrous trend is immeasurably slow because:

1. Our commitment of energy, and resources is grossly inadequate
2. Public policymaking processes for environmental management are similarly inadequate.

This paper is concerned with the air quality management process, and specifically, with the role of air quality standards in that process. In taking the rational view, we believe that it is essential for successful long-range environmental planning, but that rationality is not a substitute for a commitment of manpower and money - both are needed.

A primary objective of this paper is to broaden the scope of thinking about air quality standards by placing them in the context of a systems view of air quality management. To this end we offer prescriptions (popularly called normative models) for what the standard-setting process should consist of. (There exist many possible models of the decision process. It does not appear to be fruitful to search for an "optimum" process model.) Our approach is to visualize an ideal-type rational model and show how to approach it in the real world of poor information, uncertainty about future events, imperfect legal and administrative structures, and pressures exerted by various interest groups. This gap-closing method synthesizes a model which combines pure-rationality and reality and will, it is hoped, be suitable for real world use.*

A second objective of the paper is to derive, from the model, the information requirements of the various decision-makers who are part of the management process. It is helpful to consider Churchman's definition of information (Churchman, 1961).

> By "information" we mean recorded experience which is useful for decision-making. . .we will be considering information from the point of view of its effect on decisions that people make.

In air quality management, as in many other fields, much data are gathered but relatively little information is extracted from these data. The existence of a systems model of the air quality management process permits us to specify what data should be gathered and how it should be transformed to meet the specific information requirements of the decision-makers. Holding an ideal-type rational model in front of us also helps us to assess the magnitude of the effort and the costs of gathering data to fit the rational model. Since the quality of information is a limiting factor in the decision process, allocation decisions for research and data gathering are very important.

Rationale

Our basic value assumption for air quality management is that some balance must be reached between improving the quality of the air and the cost of doing so. We do not subscribe to the goal of "absolutely clean air" or "pollution levels where *no effects at all* are experienced." There are levels of pollution that people will tolerate because the cost of reducing effects further is considered prohibitive by a large number of them. We believe, further, that:

1. Goals for air quality should reflect the desires of the public to eliminate sensory irritation, aesthetic, health, and economic damage.

2. Air quality standards should reflect these goals, which should be compromised only after weighing the relative costs and benefits of different levels of air quality.

Reality-testing

It is necessary to add a note of qualification. Some of the activities in the proposed decision process and some of the techniques which are envisioned for the transformation of information may not be able to withstand the test of real world application, although they appear reasonable now. This is to be expected in an ideal-type model. It is, therefore, critically important for the management process to provide for feedback and evaluation, not only of air

---

*The synthesized model is similar in concept to Dror's "optimal" policymaking model, which combines rational and extrarational elements. (Dror, 1968).

quality and effect data, but also of performance data for the decision models and for organizational behavior. Conceptually, we envision an adaptive, learning model of decision-making, where the quality of the physical, behavioral, and decision models is continuously upgraded. Frequently, the identification of substandard performance will trigger a search for better methods. This process can be enhanced if the system is structured to include the proper information feedback and evaluation mechanisms. One of the harsh realities of organizational behavior, especially in a bureaucracy, is its resistance to change. However, there is reason to believe that air quality management is much more open to change than older, non-science-oriented organizations, and is capable of accepting such a change-producing model.

II. Systems Analysis and the Air Pollution Decision-Making Environment*

It is essential to our analysis to look carefully at the decision making environment of air quality management. Failure to do so usually leads to systems analyses or optimization calculations which have no link to reality, and their results have no chance of being adopted. A comparison with the environment in other areas of application may help illustrate several points.

In military and industrial settings the most important factor to the systems analyst is the existence of a central authority, which has the power to implement a preferred solution. The analyst still needs to consider physical realizability, feasibility, and constraints imposed by the environment and the customer. But once alternative designs (or policies) pass these tests the probability that they can be implemented is high. When the Department of Defense selects a weapons system and lets contracts for its development, it knows that all of the contractors will play their roles according to the rules of the game — formal and informal. These rules allow for cost overruns, delay, and degradation of quality from original specifications. However, review and bargaining procedures are built into the process to contain such "abuses" within acceptable bounds. Nevertheless, the Department of Defense has the centralized authority and the power, through reward structures, to have its decisions implemented. Generally, the same ability to implement decisions exists in private industry. When top management decides on a course of action, it has the power to implement the preferred alternative. Top management does not ordinarily worry about mobilizing support among lower executives and workers for its choice. Its authority is accepted as legitimate by all those at lower levels of the industrial organization.

E. S. Quade's (Quade, 1969) comparison of the military-industrial problem environment to the "social" problem environment is instructive:

---

*Papers by E.S. Quade (Quade, 1969) and Y. Dror (Dror, 1969) were very helpful to us in this section.

In industrial and military applications the problem is far more likely to deal with a completely man-made and directed enterprise...that was designed with a purpose in mind and has a structure that follows the laws of engineering and economics. Goals can be defined, authority is clear-cut and cooperative, and the underlying design can be discovered and modeled.

In contrast:

...an attack on problems of air pollution, urban renewal, vocational rehabilitation, or criminal justice, involves working with goals that are obscure and conflicting, where authority is diffuse and overlapping, and where the structure has grown without conscious design.

In the air pollution decision-making environment, as in many others in the social sector, authority is fragmented and power to achieve certain ends is weak. A systems analysis which assumes there is a single decision-maker, with authority to implement the preferred alternative, does not adequately reflect reality. When this happens the entire formulation of the analysis is inappropriate and it is unlikely to yield information that can be used by any of the real decision-makers.

What then, is the decision authority in air quality management and how does it affect our analysis? First, we shall look, in broad terms, at the division of responsibility (authority) among federal, state, and local air pollution agencies. We also keep in mind that authority does not automatically convey power — an agency may be authorized by law or regulation to perform a function, yet it may have inadequate power to do so.

The key metapolicy choice in air quality Management was made by the Congress in the Air Quality Act of 1967. The choice — to fragment authority between federal and state government and to assign the major role to the states — was clearly in favor of industrial interests and opposed to the wishes for a greater federal role expressed by HEW Secretary John Gardner and National Air Pollution Control Administration Director, Dr. John Middleton (O'Fallon, 1968).

The Act requires the Secretary of HEW to establish air quality regions and air quality criteria for the major pollutants; to publish techniques for controlling those pollutants and the costs of controlling them. The Department reviews the air quality standards and the plans for implementaion drawn up the the states.

The states play the key role: they are responsible for adopting standards for air quality in each designated air quality region in the state. They must also draw up a plan for implementaion which is consistent with the standards and is likely to achieve them in a reasonable length of time (Air Quality Act) Thus, as analysts, we can proceed freely with this limited problem, knowing that the states have the power to establish air quality standards.

But, the problem is not quite so simple. We see that even if we are able to predict the consequences (effects to people and property) of stated levels of air quality, we cannot predict the costs of attaining that air quality unless we assume specific technological alternatives. Thus, we cannot set standards based on a weighing of costs and benefits without assuming the technological means (control devices, process changes, fuel restrictions, location changes) needed to implement the air quality alternatives. Yet no air pollution agency has the power to implement *directly* a preferred technological alternative, which may consist of many individual control measures. We face a dilemma in our analysis: the rational approach to establishing an air quality standard depends on the assumption of technological alternatives which the air pollution agency has no power to implement. It is an unreal model of the decision-making environment that assumes such power is available. To illustrate, suppose we find that, for Los Angeles, the "cost-effective" air quality standard for oxidant is 0.05 ppm (1 hour average, not to be exceeded on more than 30 days per year), but that this result depends upon the assumption of the technological alternative: replace all internal combustion engine (ICE) vehicles by steam powered vehicles within 10 years. Clearly, the air pollution agency does not have the power to implement the preferred alternative.

We have differentiated between three levels of alternative — *air quality* (policy), *technological* (control), and *strategy* (means of implementing technological alternatives) — to aid in visualizing the fragmentation of authority and lack of power in the actual decision-making environment.

The foregoing discussion should not be interpreted to mean that the air pollution agency cannot *influence* policy. There are various strategic alternatives available to *promote* implementation of the desired technological alternatives. Returning to our hypothetical example, suppose the recommended technological alternative is to phase out all internal combustion engines in Los Angeles County. The first step might be to prohibit sale of new ICE vehicles at some specified future date. The prohibition cannot be by edict of any air pollution control agency. It can be implemented by a law banning the sale of ICE vehicles (assuming that such a law can be enforced); by a tax on ICE vehicles (which requires alternative means of transportation to be effective); or, by market mechanisms which provide attractive alternatives.

Structure of the Evaluation Model

The view that an analyst should take is that any technological alternative which he proposes may be implemented by a variety of strategic alternatives. The alternative he recommends may have a low probability of being implemented, or it may take much longer to implement than he believes it should. His analysis should attempt to incorporate the probability of implementation at specific future dates, and the decision model should give appropriate weight to the probabilities of various contingencies (unexpected future events).

It is our thesis that a rational decision-model must formulate the decision problem as one of uncertainty and weigh, as part of the *decision-criterion,* the probability of implementation of alternative policies. The probability of achieving a certain policy may be as important as the cost of achieving it in making the decision. The decision-criterion should specify how to combine the probability of occurrence with the expected costs and benefits. In a later section we shall consider some possibilities offered by mathematical decision theory for constructing such a model.*

Political Feasibility — Implementability

The importance of considering political feasibility in policymaking is suggested by Dror (Dror, 1969) and supported by Quade (Quade, 1969). Political feasibility is only one, although frequently the dominant, component of a broader concept which we label "implementability." There are organizational, technical, economic, and physical factors which may reduce the probability that an alternative will be implemented *as originally conceived* and analyzed. In estimating a probability of implementation for our decision model, all of these factors should be considered. It is clearly unrealistic to expect more than a very rough estimate of such a probability. The best hope for obtaining reasonable estimates of such hard to predict events appears to lie in the systematic processing of expert opinion, as for example, by the Delphi method.** Delphi, developed at the RAND Corporation primarily by Olaf Helmer, Normal Dalkey, and E.S. Quade, has aroused widespread interest for application in technological forecasting, corporate planning, organizational decision-making, and policy evaluation. (Quade, 1969)

Although consideration of implementability is likely to improve decisionmaking, it contains a drawback that should not be overlooked. A large part of the reason we have serious environmental quality problems today is that policymaking in the past has been incremental — patchwork responses, *dictated by what is politically feasible* — rather than long-range and goal-oriented (Atkisson, 1967). Thus, to overcome the curse of incrementalism, we shall have to reconsider feasibility constraints in the context of long-range planning. That is, the probability of implementation (or political feasibility) will have to be weighed against the requirements for goal-achievement. To illustrate in an extreme way, suppose we are threatened by extinction due to pollution, but that this threat is not believed by a large number of people. Suppose a course of action exists which can overcome this threat, but it would require a massive reallocation of resources. Such an alternative may have a very low political feasibility. Yet to those who perceive the danger of global ecological disaster, the utility of any other course of action is zero, and they

---

*In the terminology of mathematical decision theory, "uncertainty" refers to states whose probabilities are unknown. When the probability of future states is known or meaningful subjective probability estimates can be made, the situation is referred to as "risk" rather than uncertainty. We shall use the term uncertainty, in its more general and popular sense, to include the situation where the probability of future states is known.

**A detailed description of the Delphi method may be found in (Helmer, 1966).

would choose the only alternative appropriate to the situation, regardless of its probability of implementation. This suggests that combining utility measures with the probability of the outcome of contingent events − as in von Newman-Morgenstern Utility Theory* − may provide theoretical insight to the decision problem.**

We particularly like Dror's warning on this matter (Dror, 1969)

What really worries me is ... the danger that every political feasibility prediction tends to ignore the capacities of human devotion and human efforts to overcome apparently insurmountable barriers and to achieve not only the improbable but the apparently impossible. A good policy may be worth fighting for, even if its political feasibility seems to be nil, as devotion and skillful efforts may well overcome political barriers and snatch victory out of the mouth of political infeasibility.

III. Air Quality Standard Decision Model

This section focuses on the decisions relevant to setting the air quality standard. Reference is made to Fig. 1, which is a model of the decision process. In the boxes are statements which specify tasks to be performed, and decisions to be made. Performing the task may itself require a series of secondary decisions, not shown in the diagram. For many of the tasks we have a clear idea of what secondary decisions are required, but for some we do not. Some of the tasks are extremely complex − e.g. synthesize technological alternatives. These will undoubtedly require simplifying heuristics which have not yet been developed.

Goal Setting

A feature of the standard setting model is that it considers values and aspirations of the public as a basis for establishing air quality goals (not standards).† The underlying rationale is that public policy should meet the needs and desires of the public. If "amenity benefits" of pollution control are desired − e.g. elimination of visibility loss. odor − they are as valid bases for control policy as health damage and economic loss.

Since the public may not be aware of certain damage caused by or exacerbated by pollution, it may undervalue the elimination of certain effects. To counteract such undervaluation, damage information from expert sources is weighed by the decision-maker along with the public's desires.

Combining individual values into a group measure for the purpose of making a decision for that group has been a stumbling block in welfare economics. The goal of constructing a social welfare function from individual utilities is believed to be unattainable by many economists because, they claim,

---

*A detailed discussion of utility theory may be found in (Luce & Raiffa, 1957).

**The obstacles in applying utility theory to group decision-making are formidable, if not insurmountable.

† This was suggested by Arthur A. Atkisson.

Fig. 1 – Air Quality Standard-Setting Decision Model

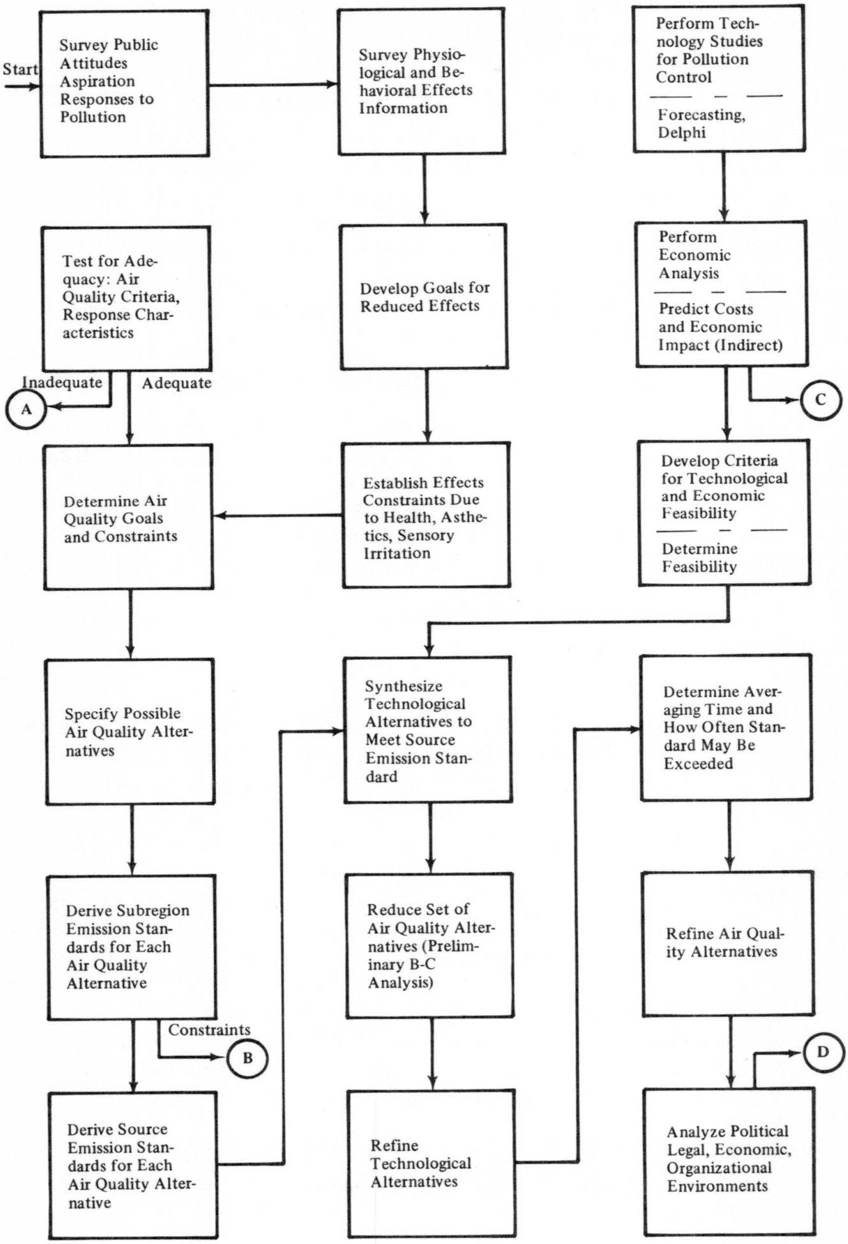

Start →

Survey Public Attitudes Aspiration Responses to Pollution

Survey Physiological and Behavioral Effects Information

Perform Technology Studies for Pollution Control

Forecasting, Delphi

Test for Adequacy: Air Quality Criteria, Response Characteristics

Develop Goals for Reduced Effects

Perform Economic Analysis

Predict Costs and Economic Impact (Indirect)

Inadequate (A) ← | Adequate

C

Determine Air Quality Goals and Constraints

Establish Effects Constraints Due to Health, Asthetics, Sensory Irritation

Develop Criteria for Technological and Economic Feasibility

Determine Feasibility

Specify Possible Air Quality Alternatives

Synthesize Technological Alternatives to Meet Source Emission Standard

Determine Averaging Time and How Often Standard May Be Exceeded

Derive Subregion Emission Standards for Each Air Quality Alternative

Reduce Set of Air Quality Alternatives (Preliminary B-C Analysis)

Refine Air Quality Alternatives

Constraints → B

D

Derive Source Emission Standards for Each Air Quality Alternative

Refine Technological Alternatives

Analyze Political Legal, Economic, Organizational Environments

32

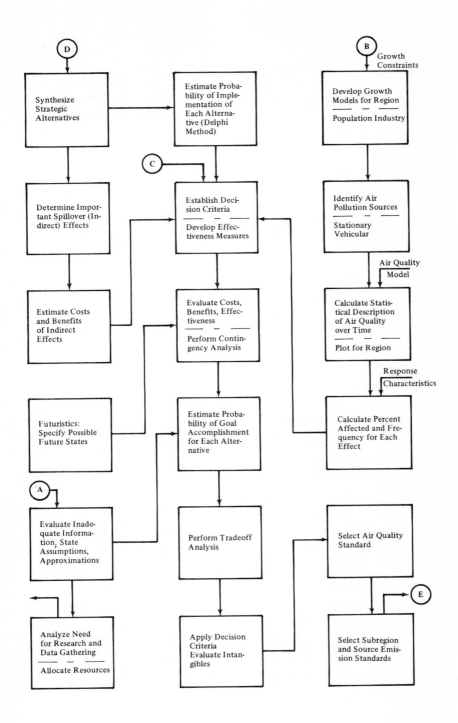

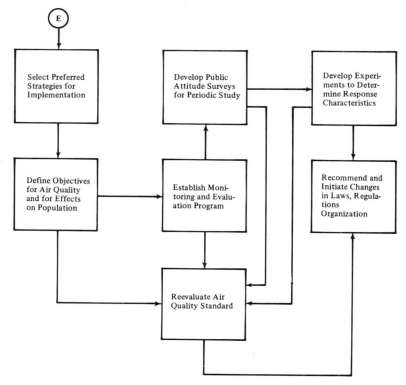

E

| Select Preferred Strategies for Implementation | Develop Public Attitude Surveys for Periodic Study | Develop Experiments to Determine Response Characteristics |

| Define Objectives for Air Quality and for Effects on Population | Establish Monitoring and Evaluation Program | Recommend and Initiate Changes in Laws, Regulations Organization |

| Reevaluate Air Quality Standard |

there is no way to compare the intensity of one person's feelings with those of another (McKean, 1968). Ackoff (Ackoff, 1963) and Churchman (Churchman, 1961) disagree, believing that a science of values is possible; that valid empirical comparison of different people's values can be made. Ackoff argues that all public policy decisions require ethical assumptions about how people *value* benefits and costs. The assumptions may be implicit in the decision criterion, but they are made, nevertheless. Traditional cost benefit decisions are based on the efficiency criterion, which seeks to maximize net benefits *to society*. It contains the ethical assumption that each individual values marginal benefits equally regardless of his wealth or income. Rather than make such a demonstrably and intuitively incorrect assumption, Ackoff recommends scientific study to obtain more realistic estimates of utility or value. It appears that little progress has been made along these lines (Dror, 1968).

For our purpose it is necessary to identify those effects which are most bothersome to the public. A survey can determine what percentage of the population of a given area is aware of an effect, how strong is the discomfort, which effects a person would most like to see eliminated, and how much money he is willing to spend. Assigning importance weights to the effects is appealing as an aid to decision-making. However, it is a precarious process which generally requires questionable assumptions. If the assumptions are forgotten, the results may be viewed with an unwarranted degree of certainty − a potentially dangerous situation.

To our knowledge, relatively little survey information, suitable for goal setting, is available. Goldsmith (Stern, 1968) reports on a survey taken in the Los Angeles basin in which people were asked whether they experience eye irritation, headache, cough, nausea, and other symptoms. If they answered yes, the survey tried to determine whether they observed any relationship between their symptoms and smog conditions. More than 50 per cent of those polled were aware of eye irritation and a much lower percentage suffered other symptoms. Other eye irritation surveys, conducted by the Los Angeles Air Pollution Control District and the Air Pollution Foundation (Air Pollution Foundation, 1954) confirm that eye irritation is very widespread on some days. A recent analysis of LA APCD data (Schwartz, 1968) shows that a surprisingly large percentage of those surveyed experienced eye irritation on "light to moderate" smog days, as the LA APCD classifies them.*

Although such information clearly establishes that smog effects are widespread, it does not tell us how strongly people feel about those effects or how they value eliminating them. We require more specific information about what effects people most want eliminated and how willing they are to support their aspirations for better air quality.

Deriving Air Quality Alternatives

To obtain air quality goals for pollutant concentration from goals for reduced *effects* of pollution requires, in theory, the information contained in air quality criteria. Ideally, the criteria state the pollutant concentration and time duration for which different effects are observed, so we can start with an effect and determine allowable concentration and duration. In practice, however, complications arise. It is extremely difficult to determine cause-effect relationships because most effects of pollution appear to be due to multiple causes. The criteria statements do not adequately reflect this complexity (Cassell, 1968). Moreover, individuals exhibit wide variations in response to pollution. In order to specify an air quality goal from an "effects" goal, a reference group must be specified (assuming we make simplifying assumptions which enable us to overcome the multiple-causation difficulty). The criteria do not adequately reflect the variability of human response to pollution − they generally indicate the conditions at which effects are first observed, i.e., when the most sensitive are affected. Consequently, it is possible to set goals to protect the health or enhance the aesthetic satisfaction of the most sensitive members of the community. However, such goals will generally be long-range and unsuitable for establishing control policy. The costs of attaining these goals may make them politically infeasible.

---

*Unfortunately, the LA APCD data was taken from a sample which varied from 4 to 25 persons − too small a number for high confidence estimates.

Thus, for the purpose of deriving standards from long-term goals, it is helpful to reduce the total set of "conceivable" alternatives to a set of "feasible" alternatives. For illustration, we envision a continuum of conceivable air quality alternatives, (Fig. 2) ranging from "no further control" on one end, C_M, to "no effects" on the other, C_O. Long range goals are determined as described, and are shown near the low end of the continuum, although some long-range goals may be near the high end or may already have been attained. When we establish a feasibility constraint for reasons of cost, technology, etc., we eliminate a set of points, $[C_O, C_{F_1}]$, from consideration as possible standards.* It is important not to establish feasibility constraints which are too strict, since in the final decision-making phase it may be necessary to make trade-off studies among the constraints and relax some of them. Therefore, we do not want to eliminate from the analysis an alternative which may eventually be chosen. It should also be kept in mind that feasibility constraints will change over time.

The definition of technological and economic feasibility may prove to be a controversial issue. Industry generally considers a pollution investment to be economically feasible when it pays for itself in recoverable waste or another form of money-saving change. A trade journal article summarizes:

> Industry, in general, has dragged its feet in achieving pollution control within its own structure system. Federal, state, and local sentiment and legislation have moved industry to take some measures to try to prevent pollution of the environment. The movement, for the most part however, has been slow, sullen, and hung up on arguments of uneconomic investment. Many companies, however, have begun to realize that pollution abatement can be good public relations and, therefore, can and should be a legitimate cost of doing business. (Chemical & Engineering News, 1969)

Technological feasibility should be viewed as meaning more than present availability. The definition which is held by aerospace contractors when bidding for advanced weapons systems or space vehicles is more appropriate. Here questions are asked about theoretical limits to physical realizability, not whether they can be built tomorrow for less than $100.

Technological and economic constraints reduce the set of conceivable air quality alternatives as shown in Fig. 2. We propose to decrease the set further by establishing another set of constraints derived from minimum requirements

---

*The notation $[C_O, C_{F_1}]$ reads: the set of all points of pollutant concentration between C_O and C_{F_1}, inclusive.

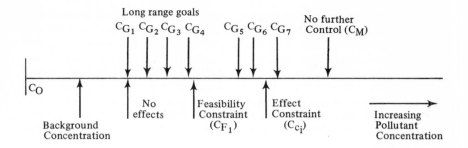

for reducing important effects of pollution. Effects are considered to be reduced when fewer people experience them, they occur less frequently for those who continue to experience them, or they are diminished in intensity. For some effects it may be difficult to establish minimum requirements for reduction; there may be no basis for selecting a percentage of the population to be protected (population of concern) and how often it can be allowed to experience the effect. Since benefits of different air quality alternatives will be evaluated in the decision phase, it is not essential to impose effect constraints where they are not clear cut. The advantage of setting effect constraints is that they reduce the work of evaluating alternatives.

Ideally, we will be able to specify several constraints, based upon different effects. Denoting these by C_{c_i}, we see that the i th constraint eliminates the set of conceivable alternatives to the right of C_{c_i} — i.e. the set $[C_{c_i}, C_M]$. If $C_{c_i} > C_{F_1}$, then the set $[C_{F_1}, C_{c_i}]$ contains the totality of air quality alternatives. If, however, $C_{c_i} < C_{F_1}$, then there are *no* possible air quality alternatives remaining after both constraints have been applied. Practically, we need to relax either the effect constraint or the feasibility constraint until there is a set of possible air quality alternatives. It may be best to analyze alternatives as though they were feasible, with an eye toward performing a trade-off analysis to relax constraints in the final evaluation phase.

Synthesis of Technological Alternatives; Subregion and Source Emission Standards

A cost-benefit evaluation for setting air quality standards cannot be performed without assuming technological (control) alternatives. We see the need for two major steps to provide a sound basis for deriving technological alternatives from the air quality standard. These are:
1. Dividing the air quality region into subregions and establishing a mass emission standard for each pollutant in each subregion.*
2. Establishing source emission standards for each source category in each subregion.

---

*The subregion emission standard was suggested by Arthur A. Atkisson.

Apart from aiding the synthesis of control alternatives the emission standards are valuable in the air quality management process. The subregion standard provides a basis for making tradeoff decisions between industrial growth in a subregion and stringency of control. Assuming that a set of subregion standards has been calculated from the air quality standard, source emission standards can be set to just meet the subregion standard or, by imposing stricter emission standards on existing sources, to leave a margin for further growth in the subregion. Zoning powers can be exercised to prevent new industry from locating in a subregion when the emission standard has been reached. It is not yet clear how uncertainty due to lack of knowledge about cause-and-effect, about pollutant transport, and other factors can best be taken into account — whether in setting the subregion standard or in setting the source emission standard to leave a wider margin for error. The source emission standard also provides a basis for enforcement while leaving the polluter free to choose his best method for abatement. The analyst must assume that he does choose the method the analyst thinks is best. No doubt, in reality, some industrial decision-makers will have a different opinion of what is best.

It appears that the preliminary analysis, which establishes constraints and narrows the list of possible alternatives, should not require subregion and source emission standards to be determined before the standard is chosen.

Establishing emission standards for each pollutant type in each subregion simplifies the job of synthesizing control alternatives. The practical value of such simplification may be very great because of the enormous complexity of the total problem. Further research is required to specify the detailed procedure for deriving subregion and source emission standards and sets of control alternatives, and to verify by simulation that the procedure yields control alternatives which are consistent with the air quality alternatives that they are designed to meet.

One of the important features of the process outlined here is that the entire decision problem is "solved" at one time, in an internally consistent fashion. The choice of an air quality standard determines subregion emission standards which determines source emission standards, which yield a preferred technological alternative and strategies for attaining it. As noted in Sec. II, the choice of a preferred strategy will not necessarily yield the desired technological alternative — there will generally be a high degree of uncertainty over implementation. It is essential to have contingency strategies available if the preferred strategy does not "perform" as expected. One of the important elements of the decision criteria is how well alternatives look under a variety of future states and how easily they can be changed to meet the unexpected.

Model for Calculating Consequences

The ideal-type model requires that the total consequences of pollution for each alternative be calculated for each year of the time-period of the analysis. A consequence is the *total occurrence* (per year) of an effect. It is calculated in natural units — e.g., number of *people-days* of eye irritation per year, days of visibility loss per year. This step precedes valuation: the process of placing a value on an effect, usually by translating its worth to dollars. Since many benefits of reduced pollution do not lend themselves to quantification in dollars, the decision-maker may want to perform a trade-off analysis or construct an effectiveness function from the consequence data.*

A shortcoming of using lumped consequence data is that it does not distinguish among individuals. It leads naturally to the assumption, criticized by Ackoff (Ackoff, 1963), that each individual values a marginal unit of benefit at the same amount. Methods for overcoming this difficulty of cost-benefit analysis should be investigated.

It is possible to obtain better quality effect data by constructing response characteristics which show the *probability distribution* for the percentage of the population that experiences a particular effect at a given pollutant concentration. This differs from the widely held view of response characteristics, where the percentage affected *at a given pollutant concentration* is assumed to be constant. Here, the percentage affected is a random variable which is described by a probability function (Schwartz, 1968). Thus, given a statistical description of pollutant concentration over time, for a given location, we can plot "percentage-frequency" curves, which show how many days per year a given percentage of the population of that area suffers from the effect under consideration. To apply this method to calculate the benefits of reduced emissions, we assume that the response characteristic (not the percentage affected) does not change. This is a plausible assumption. To illustrate, it says that under a future pollution control program, the same percentage of people will suffer from eye irritation at 0.10 ppm total oxidant as do today at 0.10 ppm total oxidant. An effects survey coupled with air quality monitoring can easily test this assumption. If it is inadequate (but not ridiculous), the response characteristics can be updated as a basis for future decision-making.

There is an important class of short-term effects which are amenable to this treatment — eye irritation, headache, shortness of breath, cough, minor respiratory illness, and possibly visibility loss if the behavioral or psychological response, not the physical measure of visibility, is considered. For the other effects listed, there is some problem in separating the contribution of pollution from the contribution of other factors which alter the physical and emotional condition of the individual. However, it should not be a difficult matter, especially for eye irritation, shortness of breath, and headache.

Ideally, we should like a predictive model of physiological and behavioral responses to pollution. Such a model of an appropriate breakdown mechanism in the body would enable us to predict, statistically, short-term and long-term consequences. Friedlander (Friedlander, 1968) proposed a model of this type. It hypothesizes that pollution hastens the natural process of aging,

---

*See (Fleischer, Atkisson, Kreditor; 1968) for a more detailed discussion of the methodology.

39

thus increasing the number of deaths in a period of high pollution. The predictive equations received promising verification using data from the London disaster of 1952. It would, however, be premature to generalize from these results.

In the absence of breakdown models which is the usual case, it is necessary to resort to statistical analysis. Although of great value in many applications, the prospect for obtaining accurate predictive models, for serious illness, be epidemiologic analysis is not good. It is assumed that statistical analysis can tell us the number of excess deaths or illnesses occuring in periods of unusually high pollution, if adequate baseline data is available. But, it cannot tell us what part of the normal death rate is due to pollution. Nor can it tell us what role pollution has played in the long-term health of an individual or by how much it has increased his susceptibility to disease.

For example, for a comparison of urban to rural lung cancer mortality data, Ridker (Ridker, 1967) concludes that *at most*, 18 per cent of the total cases can be attributed to air pollution. However, this figure might be too low if cigarette smoking is synergistic with air pollution in producing lung cancer. In that event, some of the cancer cases attributed to cigarette smoking would *not* have occurred without air pollution.

To calculate the consequences for reduced pollution alternatives, an air quality model is necessary. One component is a physical dispersion model which describes the dynamics of air mass motion for a large urban area. Such a model would be capable of generating a statistical description (probability distribution) of pollutant concentration over the region. This model is not available, nor is it being developed. Its development cost and running cost would be fairly high from the National Air Pollution Control Administration frame of reference.* Existing gaussian plume models are not sufficiently accurate.

To complete the air quality model, chemical reactions in the atmosphere must be incorporated. A model of the photochemical process, which produces oxidizing smog, is particularly needed. Although the process is enormously complex, prospects in this area appear to be good.**

In addition to response characteristics, "breakdown models," and an air quality model, the following information is required to estimate the total consequences of each alternative, over the time period of the analysis:

---

*Stanley Greenfield, in a personal communication, estimates a development cost of roughly $300,000 and computer running time of about 1 hour to simulate 1 day of real time, *on an Illiac IV computer* (fourth generation). With this running time, clever statistical approaches will be needed to generate air quality statistics — a crude Monte Carlo approach will be impossibly expensive. To help place the cost in perspective, the NAPCA may not make any grant of more than $1.5 million for research and development of new or improved devices or methods to prevent or control discharges into the air of various types of pollutants (Air Quality Act; Sec. 104 (a) (2)).

**Wayne is simulating the process in parametric form to explore the effect of different parameter estimates in approaching empirical observations. (Wayne and Earnest, 1969)

1. A growth model for the region — population and industry (by type)
2. Geographical distribution of population
3. Pollution sources, by location
4. Estimates of pollution output for each source over the period of the analysis (requires predictions of future device or process efficiency)

Evaluation of Costs and Benefits: Pure-Rational Model

Ideally, the benefit-cost analysis for the air quality alternatives proceeds in approximately the following manner:

1. First a planning horizon, which is the time period for the analysis, is selected.

2. The cost of the preferred technological alternative for each air quality alternative is estimated for each year of the planning horizon. Capital costs, maintenance, repair, and operating costs (labor, material, taxes, insurance) are considered.

3. The consequences (total number of people affected) are calculated for each effect for each year of the planning horizon. A unit cost is established for each effect, if possible.* Those consequences that cannot be converted to monetary terms are labeled "irreducibles" and are considered after all the dollar costs and benefits are calculated.

4. For each year, the consequences are multiplied by their respective unit costs to obtain a total cost. It is important to tabulate the costs and benefits by year because of the commonly exhibited preference for a benefit in the near future to an equal benefit in the more distant future. This time preference for money manifests itself in an interest charge on capital. For public policy decisions the term "discount rate" is used. There is some controversy over the proper interpretation of the discount rate in the public sector.†

5. A discount rate is chosen and the present value of future costs (and benefits) is calculated as:

---

*Economists have derived unit costs for death and illness, but they vary widely from one study to another (Klarman, 1965), (Goldman, 1967), (Ridker, 1967). Some individuals do not accept these figures for political, philosophical, and methodological reasons.

† For a review of the competing views and a recommendation for future practice, see *Economic Analysis of Public Investment Decisions: Interest Rate Policy and Discounting Analysis*, A Report of the Subcommittee on Economy in Government, Washington, Government Printing Office, 1968. The dissenting and supplementary views indicate that the issue is as much political and philosophical as economic, and will not be resolved to everybody's satisfaction.

$$\left[\begin{array}{c}\text{Present Value (PW)}\\\text{of Costs}\end{array}\right] = \sum_{j=0}^{N} \frac{C_j}{(1+i)^j}$$

Where C_j =   cost in j th year
         i =   discount rate
         N =   planning horizon

6. A decision criterion, such as maximize net present value (benefits minus costs) is selected and the preferred alternative is identified. Only then, is the effect of uncertainty on the decision considered.

We believe that this evaluation model is inappropriate to the present demands for several reasons:

1. The situation is fundamentally and overwhelmingly one of uncertainty, for reasons given in the previous section. To formulate it originally in terms of best estimates (as if they were certain) inevitably leads to insufficient emphasis on the uncertain aspects. Performance under unforeseen contingencies, which is an extremely important factor in the decision, may be overlooked.
2. Important costs of operating the air quality management system and carrying out the preferred *strategic* alternatives are not considered.
3. Spillover costs and benefits may be overlooked.
4. An important portion of the benefits is not easily quantifiable in dollar terms. These benefits should not be relegated to secondary status.
5. In spite of the fact that differences in timing of costs and benefits are taken into account by discounting, the cost-benefits "optimization" is made at one point in time. "An optimum decision, made at one point in time, is generally suboptimum in terms of subsequent times." (Miller and Starr, 1969). The cost-benefit structure does not allow us to consider the possible effect of present decisions on future choices.

Two experts on scientific management summarize their view of the quest for optimum solutions. (Miller and Starr, 1969) "It is always questionable whether the optimum procedure is to determine *the* optimum value." They elaborate:

1. A decision is generally optimum at only one point in time. "Since we are limited in our ability to foresee the future it follows that it is useless to go to extreme lengths to search for the "*most* optimum" decision. "It is complex enough to decide *how far to go*" (emphasis in original).

42

2. There are frequently an enormous number of possible choices of action. Any attempt to obtain information about all of them would be self-defeating.
3. There are virtually innumerable factors outside the control of the decision-maker. These states of nature affect the decision outcome. It would be impossible to list all of them let alone to determine the totality of their effects in order to determine the optimum action.

Evaluation: Proposed Approach

In presenting our view of the evaluation procedure, which attempts to account for important elements of reality, we do not offer a fully developed methodology — primarily because we do not have one. We indicate what information should be considered and which decision criteria may be helpful in selecting a preferred alternative.

Perhaps, the least controversial input to the evaluation process is the cost of the alternatives. In addition to capital, maintenance, and operating costs of the technological alternatives, it is important to include costs of the management process if they differ among alternatives. (The costs of implementation will vary from strategy to strategy but these may be swamped by the control costs). We should also attempt to estimate the effect on future costs of strategies that fail or are only partially successful. To carry out such a contingency analysis an exercise in "futuristics" may be helpful. (It may be of greater value in expanding the range of alternatives). (Dror, 1968). Scenarios for alternative futures in air quality management and the larger physical, social, and political environment can be prepared. Delphi experiments can be conducted to make specific predictions about technological progress, to aid in the synthesis of technological and strategic alternatives, and to predict the probability of implementation of the alternatives — a vital input to the evaluation process.*

In addition to variables affecting implementability, the following should be considered in developing different future states: growth of industry and population; and, change in the state of knowledge about the effectiveness of control methods, dispersion of pollution, chemical reactions in the atmosphere, and physiological and behavioral responses to pollution. These elements of uncertainty, along with implementability, determine the overall probability of attaining air quality goals. If we could estimate these probabilities with some confidence, they would be an important item of information in making the final selection. Although there may be no basis for expecting an accurate estimate of the probability of goal attainment, the effort to determine the effect of uncertainty should be undertaken.

On the benefit side of the comparison, we look at the reduction of the incidence of major effects. We identify the major effects from the survey of

---

*"Futuristics" is attracting widespread interest and some criticism. For example see: (Kahn and Wiener, 1967), (de Jouvenal, 1968), (Jantsch, 1967), and the periodical, *Futures*; critical reviews are (Roszak, 1969), (Fromm, 1968).

attitudes and responses of the public and physiological evidence that we use to establish air quality goals earlier in the process. We should explore weighting the effects or assigning utility values to them, although assumptions, which are not strictly justifiable, must be made. The trade-off study, which seeks to establish relative values of the effects (i.e. marginal substitution rates), is similar to weighting and suffers from the same theoretical defects.

In looking at the projected incidence of effects for some years into the future, we are tempted to ask whether discounting is applicable to physical effects as well as to dollars. Will a headache due to smog cause as much discomfort a year from today as it does today, and should it be valued the same? Normative economic theory directs us to discount future benefits, regardless of their units but we are not certain that individuals would show such time preference. The question is not merely academic because the value of the discount rate frequently affects the choice of alternatives. A high discount rate (e.g. 12%) causes the future to be valued much less than a low discount rate (e.g. 5%). Thus, the higher the discount rate, the more the decision is biased in favor of short-term benefits.

The comparison of alternatives on a small number of important effects is consistent with current practice, (Quade and Boucher, 1968) and may represent the limit of our ability to handle complexity. If one air quality alternative is superior on all effects, over time, and costs less than the others, it is a dominant solution. However, the dominant solution may have a lower probability of implementation or goal attainment than a competitor and, therefore, is not truly dominant.

If the *degree* of goal accomplishment for each alternative can be assigned a utility value, then it may be possible to derive a "payoff matrix" in terms of utility. The alternative with *maximum expected utility* which costs less than a specified maximum is chosen. The arbitrariness of the assumptions needed to fit the evaluation into this decision model may be unacceptable.

Along simular lines, the possibility of applying the theory of decision-making under risk and uncertainty should be explored. If meaningful "payoffs" can be obtained for several possible futures (states of nature, then the five or six decision criteria, commonly used in this situation, can be applied.*
Usually, a clearly superior solution is not identifiable, leaving the decision-maker to choose the criterion that appeals to him. The procedure does not aim to relieve the decision-maker from using judgment and intuiti   It hopes to sharpen these faculties by enumerating the possible outcomes under different assumptions.

In a detailed examination of the multiple-attribute decision problem, which is similar in structure to ours, MacCrimmon, (MacCrimmon, 1968) concludes that relatively simple decision criteria are necessary in a complex problem. He believes, that to evaluate several dimensions of effectiveness along

---

*Detailed discussions may be found in (Luce and Raiffa, 1957) and (Miller and Starr, 1969).

with uncertainty exceeds our bounds for processing complexity. He recommends as decision criteria *dominance, satisficing,* and *lexicography* (comparison on one dimension of effectiveness at a time, starting with the most important, until a preferred alternative is found). It may be possible to go beyond satisficing by adopting the criterion: "Choose the alternative that maximizes the probability of achieving a *desired* reduction of effects (i.e., desired effectiveness) subject to cost less than a specified maximum." If we substitute "satisfactory" for "desired" we have a satisficing criterion. In practice the line between them may be very fine although Dror thinks that it can be kept sharp (Dror, 1968).

Before reaching the decision stage two important steps can be taken. First, the major spillover (indirect, external) effects of each alternative should be identified and quantified, if possible. Alternatives which are likely to generate serious effects can be disqualified at this stage. If they are accepted at this point, the external costs and benefits should enter into the evaluation. This is a critically important step in the entire analysis because of our propensity for neglecting external effects of seemingly rational action. Pollution, traffic congestion, and many other problems which plague us today can be traced to our neglect of externalities.

Second, outcomes of the alternatives should be checked against higher-order goals for potential conflict. An alternative may impose an undue burden on a particular group, limit freedom of mobility or choice, and consequently may be unacceptable.

Specification of the Air Quality Standard

The evaluation of costs and benefits, probabilities of implementation and goal accomplishment, consequences of unforeseen events, and major indirect effects results in the selection of an air quality standard. The standard should specify, in addition to pollutant concentration,

1. Method of measuring the pollutant
2. Averaging time for the measurement
3. How often the standard may be exceeded

Items 2 and 3 should not be stated as an afterthought, as is so often the case. Averaging time should related to the major effects from which the standard is derived. Item 3 should be derived from the expected statistical distribution of pollutant concentration for a given technological alternative and the "percent-frequency" plots for the major effects. That is, given a statistical distribution of air quality (concentration), it is possible, in theory, to determine how often the standard may be exceeded from the objective for reduced effects of pollution (for one or two effects). The point we are making is that there is a statistical relationship which can, if the quality of the data is adequate, be used to determine Item 3 so that it is consistent with an objective for reduced effects. To help clarify the issue we shall look at the proposed California standard for oxidant : 0.10 ppm, not to be exceeded more than seven days in any ninety day period or on three consecutive days (California

45

Department of Public Health, 1969). Suppose that the limit: "not more than seven days in any ninety day period" is aimed at reducing eye irritation so that K per cent of the population is affected on fewer than twenty-eight days per year. The requirement "not be be exceeded on more than three consecutive days" may not be consistent with the (assumed) statistics for improving air quality and the reduced effects goal for eye irritation just stated. Looked at in a different way, which is more appropriate to the current standard-setting process, the statement of two limits on how often the standard may be exceeded *implies two different sets of technological (control) alternatives.* When a preferred technological alternative is selected as part of the standard-setting process, as in the proposed model, this problem does not occur.

Ongoing Management Process

There are several very important aspects of the air quality management process related to the air quality standard. After the standard is set, specific objectives for pollution concentration and effects should be stated for each year (for, perhaps, 10 years into the future). These objectives are obtainable from the analysis and are consistent with the assumptions for the implementation of the standards. They serve as the measure of desired system output for a feedback and change process.

It is vital that effects be monitored as well as air quality because the danger exists that the pollution control effort will be misdirected from reducing the effects of pollution, where it belongs, to meeting air quality standards. This would not pose a problem if we knew the cause-effect relationships, but we do not. The emphasis of the Air Quality Act is on setting and implementing standards. Evaluation of progress will be in terms of how closely the standard is approached. Based on a vast body of literature on organizational behavior,* it can be predicted that the efforts of the management program will be displaced from the consequences of pollution to the standards.

To guard against this danger it is essential to review the adequacy of the standard frequently. Monitoring of effects as well as air quality is essential to this goal. It is necessary to obtain effects data over fairly long periods of time to detect significant changes. A program of periodic (or perhaps random) sampling should be compared to a continuous survey of effects. In addition to evaluation, these data should be used to construct or update response characteristics to increase the accuracy of the systems model. Attitude and behavioral surveys should also be taken periodically.

An effective management process requires interaction and cooperation between federal, state, and local agencies, which may impose additional constraints on progress. Although monitoring of effects should be a local responsibility, it may have to be supported by federal aid because of a lack of resources. Similarly, the states should carry out a large scale analysis and evaluation program but may lack qualified personnel and money to do so. In

---

*For example (Katz and Kahn, 1966).

46

view of the very meager resources of state air pollution agencies, it may have been a serious mistake to have placed primary responsibility on the states. It appears likely that if adequate analysis is to precede the establishment and implementation of standards, it will have to be performed by the NAPCA or by private consultants.

To support the program, the federal government should engage in activities which are too numerous to consider here. Among these are:

1. Develop response characteristics for different effects and combination of pollutants.
2. Develop transformations for applying response characteristics (or criteria) in different environments.
3. Develop better air quality models.
4. Develop a computer program which combines population, growth, air quality, and human response models to calculate the consequences of pollution.
5. Report on technological feasibility and the economics of control in a much broader fashion than at present. Reporting only the cost of present control devices is inadequate. Maintenance, operating costs, market effects (price increases) due to limited production capability for control devices and limited fuel should be considered.
6. Evaluate information-gathering, decision-making models, and organizational structure and behavior. Recommend changes to the states.
7. Monitor rate of compliance of states with Air Quality Act; recommend changes in law, economic incentives and penalties, and other factors that promise to speed progress in attaining desirable air quality.

Constraints on Decision Making — Political

During formulation, policies and decisions in the public sector are subjected to multiple constraints. Such constraints arise from and because of the system in which they are produced. The latter, which loosely may be identified as the U.S. political system, can be characterized as being highly pluralist, specialized and disintegrated. These three characteristics which are interrelated — the effects of one exacerbating the others — either were designed into formal arrangements of the system in the eighteenth century (e.g. division and separation of powers and federalism) or have been fundamental to it since that time. The characteristics are reflected in the presence of many groups and organizations with varying goals and interests which present problems to be resolved by a given governmental jurisdiction because the larger society outside of government is unable to solve the problem, or because it is a shortcut to problem solution for someone. Such problems emerge as policy and decision initiatives.

In the following an attempt is made to summarize some common types of constraints especially from the point of view of air quality decisions.

The first category of constraints is connected with the necessity for government decision makers to aggregate the interests of large numbers of

47

people in the society. We may call this constraint the "numbers game" inasmuch as decision makers are acting for the larger polity. Thus, to some extent policy makers may be intimidated by the appearance of large numbers of persons at hearings and meetings, and by the even more vociferous behaviors of confrontation and demonstration. Most actors in and out of government are aware of this phenomenon, and may attempt to manipulate it in order to influence others. Governmental decision makers are aware of it as well, but they can never be s᠎᠎that in a given case it is not genuine sentiment which has been mobilized.

Another aspect of the "numbers game" is the influence of surrogate groups in representing functional communities. An example of the latter is a scientific and technical community associated with a particular field such as air pollution. Scientific and technical committees often are assumed to speak for, or to represent, the larger body from which they are drawn. Thus public and private groups array individuals and committees of the *cognoscenti* who would speak for their colleagues. Air quality agencies often appoint technical advisory committees, as do associations of manufacturers, processors, etc.

A final dimension of the "numbers game" is the representation of local or constituent interests by representatives, especially where political jurisdictions are involved. Here the legislator is torn especially between the dilemma of broader versus parochial interests. A variation on this theme may be seen in the phenomenon of political partisanship, where various forms of "arm twisting" are applied to political party coreligionists to aggregate voting strength.

A second type of constraint – public opinion – is related to the first category. Public opinion may constrain governmental decision makers in a variety of ways; most importantly, it can produce vocal, activists groups which directly articulate demands upon decision centers. Needless to say, important actors in the political process attempt to influence public opinion for their own interests through media of communication. Public like-mindeness on an issue may cause agencies to modify or postpone rational action that they might otherwise have taken until credibility with the public may be established through certain preliminary activities.

A third type of decision constraint stems from an expectation that rigorous procedures of justification and analysis will preceed any major policy deviation or new undertaking on the part of government. Such procedures not only are demanded by potentially affected groups, but by watchdogs of the system in general such as taxpayer associations, grand juries, citizen economy and efficiency committees, etc., as well. Accordingly, political decision makers may be counted upon to hold hearings, provide procedures for appeals and in general to allow airing of all sides to a question. Avoidance of such arrangements is likely to produce taxpayer suits and related litigation. The observation has been made that these types of constraints produce reluctance in government bureaucracies to enter new fields. Certainly, one may document the step by step movement into a new activity area by government – usually preceeded by numerous special studies, advisory committee reports, and national, state and local conferences – propelled by the push of special interest groups and the pull of outstanding legislative leaders. Thus, aside from unwillingness to take strident action which breaks sharply with precedent, lapse of time is a principal outcome of this constraint.

Fourth, decision making is constrained by bureaucratic priorities and, fifth, by availability of resources. The dependent and independent variable status of these two factors varies. At times organizational predisposition conditions the supply of resources available, as for example when agency survival is at stake. However, availability of resources, or more precisely unwillingness to allocate significant sums of money, more often dictates program direction and emphasis. An entire orientation in problem solution may be guided by quantity of time, staff, money, etc., allocated to problem solution, with the result that blocks of alternatives may be omitted from consideration. A suggested, albeit obvious, lesson from attainment in lunar exploration is that technical solutions to problems such as those of environmental quality probably are a function of amount of resources applied.

Constraints on Decision Making — Organizational

A major constraint on goal-oriented decision-making is provided by the criteria by which the decision-maker is evaluated. Several important works on organizational behavior — e.g. (Katz and Kahn, 1966) — have stressed that the criteria by which people in an organization are rewarded is a major determinant of organizational behavior. There is nothing startling about this observation — it merely says that people tend to be somewhat rational in satisfying personal goals. If we view the decision-maker as a man who chooses courses of action which maximize net benefits *to society*, we shall be sadly disappointed. McKean (McKean, 1968) and Downs (Downs, 1957), among others, present a more realistic model of decision-maker behavior. It views the decision-maker as maximizing a *personal utility function,* composed of such variables as salary, status, power, security, organizational health *and* service to society. Swalm (Swalm, 1966) obtained impressive empirical evidence that such a model is valid by constructing cardinal utility curves for a large number of managers at a major U.S. corporation. Most of the managers showed an extreme aversion to risk-taking on decisions which involved miniscule sums of money in comparison to the annual profits of the company.*

Typically, in government bureaus conformity to rules and predictability of performance characterize organizational behavior. Resistance to change is strong. "High payoff alternatives associated with a relatively high probability of failure are more risky to him [the decision-maker] than alternatives with a much lower payoff and a high probability of success." (Archibald and Hoffman, 1969).

---

*The decisions involved tens of thousands of dollars compared to annual profits of several hundred million dollars.

Another potential impediment to the use of scientific management techniques is the lack of knowledge about them by existing employees. They, rightly, fear loss of status and importance if new people are brought in to apply the new techniques. The personal utility-maximizing model predicts change-resisting behavior in such a situation.

It is clear that to gain acceptance of methods which are long-term goal-oriented in nature and involve techniques which are unfamiliar to most employees, several organizational changes are required. It will be essential to change evaluation criteria from a short-term, means orientation − e.g. number of applications processed, number of smoke violations issued − to a longer-term goal orientation. This generally means eliminating precise numerical yardsticks and substituting judgment. The bureaucratic administrator may be loathe to accept such a change because it entails greater risk. He could be compensated for the greater risk. This requires altering reward structures, which are highly unsymmetrical in government agencies − high-position personnel are grossly underpaid in relation to the apparent responsibility they have (Archibald and Hoffman, 1969). Such asymmetry is reasonable, however, if high-level decision-makers are so well guided by rules that they do not incur much more risk than lower-level people.

To overcome resistance to change because of inadequacy in applying new methods, upgrading the education of existing employees and selecting new employees with the required education is an obvious solution. There are barriers to carrying out such a simple solution, which we shall not delve into.

References

1. Ackoff, Russell. "Toward Quantitative Evaluation of Urban Services," *Public Expenditure Decisions in the Urban Community.* Schaller, Howard G. (ed); Washington, D.C.: Resources for the Future, 1963.

2. Air Pollution Foundation. *An Aerometric Survey of the Los Angeles Basin.* Los Angeles, 1955.

3. Archibald, R.W. and Hoffman, R.B. *Introducing Technological Change in a Bureaucratic Structure.* RAND Paper P-4025, Santa Monica, 1969.

4. Atkisson, Arthur A. "Urban Ecology: The New Challenge", *Archives of Environmental Health,* October, 1967.

5. Atkisson, Arthur A., and Bowerman, Frank R. "Urban Systems as a Framework for Environmental Waste Management Planning," Paper presented at the 40th Annual Conference of the Water Pollution Control Federation: New York, October 11, 1967.

6. California State Department of Public Health, *Recommended Ambient Air Quality Standards*, (mimeo), Berkeley: 1969.

7. Cassell, Eric J. "The Health Effects of Air Pollution and Their Implications for Control," *Law and Contemporary Problems,* XXXIII, No. 2 (Spring, 1968), 197-217.

8. Churchman, C. West. *Prediction and Optimal Decision.* Englewood Cliffs, New Jersey: Prentice Hall, 1961.

9. Downs, Anthony. *An Economic Theory of Democracy.* New York: Harper, 1957.

10. Dror, Yehezkel. *Public Policymaking Reexamined.* San Francisco: Chandler Publishing Co., 1968.

11. Dror, Yehezkel. *The Prediction of Political Feasibility.* Santa Monica: RAND Paper P-4044, April, 1969.

12. Fleischer, G.A., Kreditor, A., and Atkisson, A.A. *The Systems Concept: Its Relevance to Vehicular Contamination Control.* Paper presented at the Workshop on Vehicle Contamination Control, Los Angeles, University of Southern California, January, 1968.

13. Friedlander, S.K. "A Theoretical Model for the Effect of An Acute Air Pollution Episode on a Human Population," *Environmental Science and Technology,* II, No. 12, (December, 1968), 1101-1108.

14. Fromm, Erich. *The Revolution of Hope: Toward a Humanized Technology.* New York: Harper & Row Publishers, 1968.

15. Goldman, Thomas A. (ed.) *Cost Effectiveness Analysis: New Approaches in Decision-Making.* New York: Frederick Praeger, 1967.

17. Helmer, Olaf. *Social Technology.* New York: Basic Books, 1966.

18. Jantsch, Erich. *Technological Forecasting in Perspective.* Paris: OECD, 1967.

19. Jouvenal, Bertrand de. *The Art of Conjecture.* New York: Basic Books, 1967.

20. Kahn, Herman, and Weiner, Anthony J. *The Year 2000.* New York, The MacMillan Company, 1967.

21. Katz, Daniel and Kahn, Robert L. *The Social Psychology of Organizations.* New York: John Wiley and Sons, 1966.

22. Klarman, Herbert. *The Economics of Health.* New York: Columbia University Press, 1965.

23. Luce, R. Duncan and Raiffa, Howard. *Games and Decisions.* New York: John Wiley and Sons, 1957.

24. McKean, Roland. *Public Spending.* New York: McGraw-Hill Book Co., 1968.

25. MacCrimmon, K.R. *Decisionmaking Among Multiple-Attribute Alternatives: A Survey and Consolidated Approach.* Santa Monica: RAND Memorandum RM-4823, 1968.

26. Miller, David W. and Starr, Martin K. *Executive Decisions and Operations Research.* 2nd ed. Englewood Cliffs, New Jersey: Prentice-Hall, 1969.

27. O'Fallon, John E. "Deficiencies in the Air Quality Act of 1967," *Law and Contemporary Problems.* XXXIII, No. 2, (Spring, 1968), 275-296.

28. "Pollution: Causes, Costs, Controls," *Chemical and Engineering News*, June 9, 1969, 33-64.

29. Quade, E.S. and Boucher, W.I. (eds.). *Systems Analysis and Policy Planning Applications in Defense.* New York: American Elsevier Publishing Co., 1968.

30. Quade, E.S. *The Systems Approach and Public Policy.* Santa Monica: RAND Paper P-4053, March, 1969.

31. Ridker, Ronald G. *Economic Costs of Air Pollution.* New York: Frederick A. Praeger, 1967.

32. Roszak, Theodore. "Technocracy: Despotism of Beneficient Expertise," *The Nation,* (September, 1969).

33. Schwartz, Seymour. *A New Approach to the Analysis of Air Pollution Response Data.* (Unpublished Research), Los Angeles: University of Southern California, 1968.

34. Stern, Arthur C. (ed.). *Air Pollution.* New York: Academic Press, 1968.

35. Swalm, Ralph O. "Utility Theory – Insights Into Risk Taking," *Harvard Business Review.* XLIV, No. 6, (November-December, 1966), 123-136.

36. Wayne, L. G. and Earnest, T. E. "Photochemical Smog, Simulated by Computer," presented at the 62nd Annual Meeting of the Air Pollution Control Association, New York, June 1969.

TYPES, RANGES, AND METHODS FOR CLASSIFYING HUMAN BEHAVIORAL RESPONSES TO AIR POLLUTION

By Ronald O. Loveridge
University of California, Riverside

In the development of air quality standards, the personal responses of the American people have been singularly and mistakenly ignored. The principal control criteria have been premised on laboratory demonstrated relationships between pollutants and their biological and physical effects. Yet if the purpose of standards is to "promote the public health and welfare and the productive capacity of its (United States) population," the public's responses to polluted air should not be excluded from the formulation of control policies. It is quite possible, as Ido De Groot points out, "that the cost of air pollution as measured by disease and death and economics is far less important in the end than the loss of love and joy of life itself."[1] This paper will contend that certain health and welfare criteria as defined by the public — such to protect and enhance the pursuit of life, liberty, and happiness — should be included in the development of air quality standards.

Two objections are raised to including the personal effects of air pollution as air quality criteria. One says that the views of the public are represented in the policy process and thus the standards adopted reflect the public's dominant convictions and interests. For example, Jean Schueneman writes, "The air resources of any area will be managed — or neglected — to the extent its citizens desire and demand."[2] Available research would indicate that this position is largely a pleasant fiction of our democratic rhetoric. Nowhere does a one-to-one linkage exist between public opinion and policy decisions. And especially on matters of pollution control, the policy process tends to be dominated by special interests to the point that the health and general welfare concerns of the public receive at best an inadequate hearing.[3]

The second objection explains that while perhaps important, personal responses consist of highly subjective attitudes and behavior that are all but impossible to measure scientifically. Personal responses are thus dismissed as "intangibles" that cannot be included in the formulation of control standards. Public opinion research is, however, much more advanced than the titillating press releases from Gallup and Harris might suggest. There are — or could be — available instruments for collecting reliable and valid measures of the personal effects and interpretations of air pollution. Therefore, I propose regional pollution surveys designed to measure and evaluate responses by those who must live, work, and play in air quality regions.

Setting Air Quality Standards

The formal procedures for setting regional air quality standards have been established by the Air Quality Act of 1967. As to major steps, HEW designates specific air quality regions. The Department issues criteria documents describing the harmful effects of a pollutant or group of pollutants. Then it becomes the responsibility of the State(s) to develop air quality standards and plans for putting them into effect and enforcing them.

Air quality regions have been announced. Two criteria documents have been issued. At least several regions have submitted air quality standards for review. If the process for setting air quality standards is so explicit, why then the conference mandate to discuss the development of improved methods for setting standards? Among others, two critical problems summon attention: statements of air criteria and adoption and enforcement of standards. At this time, I want only to cite the present emphases of the criteria documents — standards will be dealt with later as policy making problems.

The two criteria documents, particulate matter and sulfur oxides, are evaluative compendiums of present scientific knowledge on the extent to which pollutants are harmful to health and damaging to property. I am in no position to critique these documents; yet, because of their importance in setting the parameters of air quality standards, I would ask whether the definitions of public health and welfare that underlie the criteria documents should not be expanded to include the personal responses of man to a polluted environment.[4]

The conclusions of the two criteria documents include the effects of the pollutants on health, materials, vegetation, and visibility. The particulate matter document also adds the effects on public concern. Nevertheless, a close reading of these documents suggests that public health effects provide the major impetus for control policies. The specifics include problems of mortality, accentuation of symptoms, frequency and severity of respiratory diseases, illness rates, and so forth that can be found at various levels of pollutant concentrations. While these concerns can be applauded, I find the implicit definition of public health too restrictive and prefer instead a more positive statement such as that proposed in the preamble of the World Health Organization: "Health is a state of complete physical, mental, and social well being, and not merely the absence of disease or infirmity." The current definition seems to focus precisely on the presence of disease or infirmity and exclude matters of the character and quality of life.

For example, the Public Health Service has indicated that environmental quality would be good enough when:

1. The health of even sensitive or susceptible segments of the population would not be adversely affected;

2. Concentrations of pollutants would not cause an annoyance, such as the sensation of unpleasant tastes or odors;

3. Damage to animals, ornamental plants, forests, and agricultural crops would not occur;

54

4. Visibility would not be significantly reduced;

5. Metals would not be corroded and other materials would not be damaged;

6. Fabrics would not be soiled, deteriorated, or their colors affected;

7. Natural scenery would not be obscured.[5]

It is these kinds of issues that I find lacking in the criteria documents. Admittedly, as one document points out, "Unfortunately, very little work has been done in this particular area of measuring the response of the public to the *nuisance* (my emphasis) of air pollution."[6] However, I propose to explore certain human behavioral responses that merit attention and that can be scientifically studied.

Personal Responses to Air Pollution

While air pollution specialists agree on general ways to examine the effects of air pollution, personal responses have largely been excluded from their attention or research. Few specialists would quarrel with Leslie Chambers' five types of effects: visibility reduction, material change, agricultural damage, physiological effects on man and domestic animals, and psychological effects.[7] Chambers' commentary on psychological effects — the major type under which personal responses could be subsumed — illustrates how narrowly such effects are commonly defined:

> *Since fear is a recognizable element in public reactions to air pollution, the psychological aspects of the phenomenon cannot be ignored. Psychosomatic illnesses are possibly related to inadequate knowledge of a publicized threat. Little effort has been directed toward evaluation of such impacts in relation to general mental health of affected groups, or determination of their role in individual neuroses. Only in practical politics has any significant action been based on recognition of the psychological attitudes induced by periodic public exposure to an air borne threat.[8]*

This paragraph, if I understand it, limits personal effects to psychopathology rather than how, for example, perceptions of air pollution change man's attitudes and actions, viz., disrupt life, work, and play habits. Yet no matter how defined, research on "psychological effects" has been scarce indeed; for despite an extensive literature search, few scientific studies on human behavioral responses to air pollution can be found.

Nevertheless, we do know that people living in polluted areas are aware of and concerned about air pollution. What we do not know is how perceptions of air pollution affect their attitudes and behavior. In an excellent review of public attitudes toward air pollution, Ido De Groot stresses a similar view, ". . .we whould know how perceived dangers of air pollution alter man's life style and belief systems, and psychological makeup."[9] Therefore, I propose to explore four general types of individual adjustments to air pollution: *psychological, social, economic,* and *political.* These classifications direct attention to major personal responses that should be studied.

It would be less than candid not to recognize that the response types pose complex, difficult, and occasionally near intractable problems of definition and measurement. In many ways these problems are those of social science, for the study of man does not result in cumulative laws of human behavior. Rather, knowledge is at best tentative and incomplete, subject to the realities of a particular historical situation. Moreover, man's attitudes and behavior are seldom measurable in equal terms, but represents apples and pears that must be compared, measured, and weighed. Perhaps, as the President's Science Advisory Committee Report on Environmental Pollution observed, "The subtlety and complexity of the problems and their quantitative nature can challenge behavioral science to sharpen its techniques of measurement and analysis."[10] Yet the problems are exaggerated because the literature on personal responses to air pollution, or other environmental problems for that matter, is limited if not nonexistent. As a result, I will primarily comment on the nature of the four personal responses to air pollution.

Psychological responses are probably the most difficult to specify, for they refer to attitudes about the impact of air pollution on the private self and perceptions of its influence on the nature and character of the human condition. Ronald Riker refers to these responses as "psychic costs" and says they are, though overlooked, of major importance:

> *This category includes everything from the anguish of death to the disappointment felt when one's view of the mountains is obscured by smog. Economists have generally ignored this category on the grounds that it cannot be accurately measured, and that in many cases its inclusion would alter the decision that benefit-cost analysis which is adequate in other respects leads to. The difficulty when considering air pollution is that there are important cases where this category would make a difference in the policy decision if it were included. Indeed, I suspect that the increased demand for clean air over the last fifty years or so comes mainly from a desire for a more beautiful environment and only secondarily from an increased knowledge of the detrimental effects of pollution.*[11]

Social responses refer to the effects of air pollution on an individual's life style. The individual can ignore its presence, tolerate its effects, or change his social habits. Various levels of pollutants could influence, for example, satisfaction with and choice of place of residence, job, recreation, vacation, and so forth. Needless to say, opinions and decisions related to these activities are central to enjoyment of life and the overall demand for a decent environment.

Economic responses focus on remedial action available through the market place. Two kinds of responses can be distinguished. One is to avoid the effects of air pollution, e.g., purchase air filter equipment or pollution resistant products. The second is to treat the effects, e.g., purchase medication or special cleaning preparations. These responses are man's effort to utilize the bounty of America's private goods to adjust to the effects of air pollution. Beyond decisions to use one's purchasing power, other factors are also crucial to evaluating economic responses, for example, ability to pay, awareness of market choices, and so forth.

And, finally, political responses refer to an individual's knowledge, concern, and policy opinions toward air pollution and its control. Does he recognize the problem? Is he concerned? What action has he taken? What does he believe should be done? (These questions will be examined in detail later.)

Personal Responses and the Policy Process

Democratic precepts and the need to control air pollution premise the endorsement of survey research on personal responses. Until policy councils evaluate personal responses, the adoption of comprehensive control policies will be delayed, perhaps thwarted. For, unfortunately, the present policy process frequently, sometimes resolutely, works against the growing public agreement on the need for clean air.

The procedures for setting air quality standards are ensconced in the democratic rules of the game, American style. The federal government will not impose uniform standards or controls; rather, ". . .the prevention and control of air pollution at its source is the primary responsibility of State and local governments." In effect, States become the primary architects of control standards, limited only by the caveats that standards should be "consistent" with the criteria documents and implemented in a "reasonable time." Further, legislative rules require that public hearings be conducted prior to the adoption of air quality standards. "This is to ensure," says John Middleton, "that affected citizens have a say as to the quality of the air they want":

> 'Participatory democracy' is Senator Edmund Muskie's pharse for this type of community decision-making. The various considerations outlined in the criteria document should be openly discussed at such hearings. The determination as to whether or not the expressed views of affected parties have been adequately considered will be based in part on examination of the transcripts of such public hearings on air quality standards.[12]

Though such "democratic procedures" would seem to mask the parcelling out to private interests the power to determine control policies, many politicians and specialists see the invisible hand of public attitudes working toward feasible and necessary controls. A.J. Haagen-Smit, for example, concludes, "Air pollution control is always a balance between the desire of obtaining as clean air as possible and the price the community is willing to pay for reaching this goal."[13] This interpretation, however, is contradicted by the casual hearing personal responses receive at the setting and implementing of control standards. Thus, one major way to change the present focus from private to public priorities is to survey those who live in air quality regions and place the results on the public record.

This call for regional surveys is grounded on three "facts of life" of the policy process. First, the citizen as citizen is largely an impotent partisan, for he lacks the resources and expertise to gain access for influence. As Dorwin Cartwright aptly explained, "Surely, the citizen who doesn't belong to a strong organization with a lobby in Washington or who doesn't know a Congressman's wife has little chance of influencing governmental policies and procedures except through rare visits to the voting booth."[14] Even if highly motivated, most citizens lack the wit and the means to make the case for strict air quality standards. And even if successfully organized into protest groups, the vocal citizen is dismissed as unrepresentative and unrealistic — one official spokesman, for instance, recently referred to the public hearings as a "mere yelling and screaming process." Also, most pollution protest groups have been noticeably ineffective in sustaining a concerted partisan influence on control standards and programs.

Second, setting standards should be viewed as a process of political bargaining among competing interests — HEW is only one partisan among many. Some political observers contend that the pulling and hauling among private interests are sufficient due process for decisions on public goods. As Robert Dahl has explained:

> For it (American polity) is a markedly decentralized system. Decisions are made by endless bargaining; perhaps in no other national political system in the world is bargaining so basic a component of the political process. . . .(yet) with all its defects, it does nonetheless provide a high probability that any active and legitimate group will make itself heard effectively at some stage in the process of decision. This is no mean thing in a political system.[15]

Yet on pollution matters, the contest among organized interests vanquishes the public, for in no way can citizens match the organization, resources, or influence of America's corporate polluters. Political bargaining has in point of fact savaged most comprehensive control standards and programs.

The results of interest group bargaining are well illustrated in a recent article by Jane Stein, "Priorities in Pollution: The SST and the Smogless Car." She writes that the SST (Supersonic Transport) has high priority and abundant funding, though it would serve about five per cent of the population, while the non-polluting car program has no priority and virtually no funds, though it would benefit almost everyone in the country:

> Although noise pollution from the inescapable sonic booms could create the most widespread environmental blight in the nation's history, the United States government has already allocated more than half a billion dollars for the SST program and

may end up spending between $1.3 and $4.5 billion to see the venture through.

. . .a different kind of environmental blight — air pollution — continues to grow in scope and intensity. America's 97 million motor vehicles. . .are producing more than 60 percent of that pollution. Yet the Senate has not even acted on legislation to spend a comparatively mere $3.5 million toward the development of automobiles that do not pollute the air. [16]

The reasons for such "resource allocation" can largely be explained by the respective influences and interests of the aircraft and automobile industries. If committed to the process of political bargaining, pollution issues will usually be won in legislative battle by the corporate interests and lost by the public.

And third, the implementation of control standards tends to provide reassuring symbols for the public and concrete benefits for organized interests. Control agencies have few friends; instead they deal almost exclusively with clientele groups who they are set up to control — namely, those industries which manage the economic resources of the region. Yet control agencies can survive only so long as they continue to secure the support of politically effective groups and through these groups to secure legislative and executive support.[17] Murray Edelman summarizes the results common to regulatory agencies:

> *1) Tangible resources and benefits are frequently not distributed to unorganized political group interests as promised in regulatory statues and the propaganda attending their enactment.*
>
> *2) When it does happen, the deprived groups often display little tendency to protest or to assert their awareness of the deprivation.*
>
> *3) The most intensive dissemination of symbols attends the enactment of legislation which is most meaningless in its effects upon resource allocation.* [18]

He further points out, "The laws may be repealed in effect by administrative policy, budgetary starvation, or other little publicized means."[19] There is no triumph for the public good when air quality standards are adopted, for the administrative process is subject to the same bargaining politics as the legislative process. And here the public has even less influence on what policy decisions are made.

The three facts of political life have two noteworthy consequences for efforts to control air pollution. For one, public priorities atrophy. The public tends to be shut out, first at the legislative, then at the accountability phase. And, second, the impact is a conservative influence on standards and programs. There is an established tendency for group-agency relationships to be highly resistant to "disturbing changes." As Theodore Lowi has observed, "As programs are split off and allowed to establish self-governing relations with clientele groups, professional norms usually spring up, governing the proper ways of doing things. These rules-of-the-game heavily weight access and power in favor of the established interests. . . ."[20] In other words, personal responses to air pollution do not effectively enter the policy arena when standards are adopted or when controls are administered.

State and local officials have little information on personal responses except anecdotes, visceral feelings, and subjective inspections of the mass media. Survey data on personal responses could be placed on the official docket to join the views of private interests and government technocrats. This recommendation is certainly not a radical step. No major politician will begin a campaign or announce his platform before polling his constituents. Nor will a major corporation market a new product without consumer research. The innovation is that surveys will be specifically designed for use in the formal deliberations of setting and implementing air quality standards.

Information on personal responses should have several major consequences. First, in an important sense, democratic rhetoric and policy practice can be consummated, for the public will be given at least an indirect hearing in the setting and implementing of standards. Until now, the public's perception of air pollution and the influence such perception has on attitudes and behavior have received short shrift in most policy councils. And second, surveys can identify and emphasize personal effects that are not presently evaluated. This data should contribute to the assessment and development of air pollution control programs. As Jan Schusky and colleagues observed:

> *Public opinion surveys are particularly valuable. . .with respect to those aspects of air pollution which are not presently capable of standardized, objective measurement. Thus, the extent to which odors, visibility effects, esthetics, soiling, and sensory and upper respiratory irritation become a "problem" is highly dependent upon the extent to which people see it as such, and are concerned or bothered about it. Further, the extent to which air pollution is a problem may also be reflected in the ways in which people alter their behavior in response to it. Thus, air pollution may be a factor in causing people to avoid downtown shopping areas, or certain recreational areas, or to change their residence or place of work.*[21]

Finally, surveys of personal responses should identify the extent to which the public demands and will support pollution control measures. This reason is perhaps the most important of the three, for present evidence suggests that the public is more predisposed to control air pollution than state and local authorities. Recent polls have found pollution control at or near the top of priorities the public believes government should do something about. For example, a recent poll of California voters found that out of five issues, voters assigned highest priority to "stopping pollution of our environment."[22] If carefully documented, the urgency and intensity of the public's cry for clean air should present a charge to policy makers that cannot be dismissed or ignored.

Political Responses: Some Specifics

While exhorting the study (and importance) of personal responses, I have been unfortunately vague on identifying specific responses. Vagueness cannot be attributed to survey research, for its skills are now sophisticated, widely known, and highly scientific. Rather, there is a striking lack of relevant conceptual or empirical work on personal responses to air pollution — hard facts and completed studies simply do not exist. I do not know, in many cases, what specific responses are significant or what measures are possible and objective. Therefore, to illustrate the kinds of responses that can be studied, selected political responses will be discussed.

Political responses refer to attitudes and behavior (whether actual or latent) toward the control of air pollution. The literature on political participation says that political involvement, even in its most elementary forms, generally depends on at least three conditions being fulfilled:

1. The problem must be recognized.
2. It must arouse some strong feelings of intesity.
3. It must be accompanied by a perception that political answers are available and that political involvement could make a difference. [23]

These conditions for political involvement provide a classification scheme to examine political responses. Thus, I will take up knowledge, feeling, and policy views toward the control of air pollution.

Political Knowledge

Political attitudes toward air pollution can be divided and explored in many ways. When we focus on data immediately relevant to the setting of air quality standards, two dimensions seem important: first, the recognition of air pollution as a problem, and second, the perception of the personal effects of air pollution. The recognition of air pollution as a problem is the most fully studied of any question on personal responses. Whether a national, metropolitan, or community pollution survey, respondents are always asked — in some manner — if air pollution is a problem for their area of residence. The results show an increasing and widespread awareness of air pollution and a positive correlation between the levels of pollution and awareness of air pollution as a problem. If carefully constructed, a measure of awareness could be used to develop standards such as W.W. Stalker and Charles Robison propose:

For a control program to be worthwhile, it would seem that concentrations of pollutants presenting a nuisance to as many as 50% of the people should be permitted for only short periods of time if at all. [24]

While presenting difficult problems of measurement, perceptions about the personal effects of air pollution also merit close evaluation. For the effects people can identify explain why air pollution is thought of as a problem. The

61

content of these responses, by indicating what people find objectionable, irritating, or dangerous, should be instructive to those responsible for setting and implementing air quality standards. Personal effects represent, however, a wide-ranging catalogue of possible complaints; to illustrate, let us take one example, loss of visibility. If a high percentage of the residents of a region are distressed, perhaps angered by the reduction of visibility, the visibility statements in the criteria documents should become more important in air quality deliberations. Moreover, a focus on visibility could have two major advantages: one, visibility is an objective measure on which we have considerable data and two, visibility could probably serve as a satisfactory indicator of other effects, many of which are premised on what people can see.

Feelings About Air Pollution

An Interlandi cartoon appearing in the *Los Angeles Times* pictured a sedate, middle-aged couple confronting their activist son who, with picket sign in hand, was off to another protest. With a sense of bewilderment and resignation, the parents ask, "Why war? Couldn't you protest against smog? Everyone's against air pollution." While it is surely true that most people oppose air pollution, the cartoon calls attention to the second major kind of political response: relative concern that air pollution should be controlled. The public faces a plethora of problems all competing for attention and resources. The question becomes how much they value the control of air pollution when compared to other private decisions and policy choices. Relative intensity of personal concern for pollution abatement should become a part of the policy process, for the public's priorities in addition to specific group pressures should be counted when air quality standards are adopted and enforced.

Despite serious problems posed by comparability and intensity, personal feelings about the control of air pollution can be evaluated by survey research. Several general measures suggest ways to begin. In a 1967 national sample, Louis Harris, for instance, asked whether seventeen federal programs should be expanded, kept as is, or cut back. First on the list of five to be expanded — including, among others, federal scholarships for needy college students and medicare for the aged — was the program to curb air pollution.[25] In a 1968 national Gallup poll, a different measure of intensity was used to sample public opinion on the effects of environmental deterioration: "About half (51 per cent) of all persons interviewed said they are 'deeply concerned' about the effects of air and water pollution, soil erosion, and destruction of wildlife. About one-third (35 per cent) said they are 'somewhat concerned.' Only 12 per cent said they are 'not very concerned.' "[26] And finally, an ingenious example of a more complex instrument to measure personal concern was developed by two students in the 1968 Cal Tech summer project, James Beck and Marianna Stapel; they attempt to measure "relationships between values, those relating to transportation and air pollution, and to provide meaningful uses for the information obtained."[27] Thus, while much work has to be done, objective measures of the public's relative concern for the control of air pollution can be developed.

Policy Views

Two kinds of policy views can be explored: orientation and policy stand. Orientation refers to how the individual directs himself action-wise. The focus would be on how the respondents have approached the control of air pollution. While policy stand is the preference for collective action: what should the "authorities" do about air pollution? As a former California legislator active in air pollution legislation reminds us, "Those of us who make the final legislative decisions often find ourselves cought in the dilemma of seeing the problem and knowing of the solutions but lacking sufficient popular support to carry them out."[28] Measures of policy views should provide exact indications of the kind and extent of the support that exists for air pollution control procedures and objectives.

The concept of orientation directs our attention to the public's political reaction to air pollution. In what manner or form do people participate? In a HEW reprint, Stephen Ayres concluded a speech with the plea: "The issues are clear cut – the opposing forces are grouping and sharpening their weapons. What is the citizen's role? It is simply this. He must stand up, be counted, and shout in a loud, clear voice – I demand pure air."[29] In spite of such democratic rhetoric, political participation is drastically limited. Even during presidential campaigns, most citizens do little more than vote – and millions do not even take this chance to participate. The literature on political participation indicates that most Americans are not well informed, not deeply involved, and not particularly active on political matters.[30] If defined as talking to friends and neighbors, writing letters, signing petitions, making complaints, advocating reforms, joining groups, and the like, political involvement for the control of air pollution is also limited to small numbers of people. Nonetheless, one useful measure would be a comparison of personal political activity on air pollution with other environmental and social problems. It could indicate that the participation rate is noticeably higher on air pollution matters than on most other problems in the specific air quality region. And there is evidence to suggest that political interest in air pollution control is increasing, both in individual actions and in protest and professional group activities.

It would be a misinterpretation to equate participation rates with political support, for most people do not know what can or should be done to control air pollution. As Gilbert Seldes pointedly explained to the 1962 National Conference on Air Pollution, "Everyone agrees on the facts – everyone says the situation is intolerable. And with the exception of yourselves and a few friends, everyone seems to feel that nothing can be done about it."[31] In all available studies, this feeling of political impotence is clear. People do not participate because they do not see that political alternatives are available or that political involvement could make a difference. To illustrate, let me quote the conclusions from an exploratory survey I conducted with 45 residents of San Bernardino, California:

*Air pollution is not a politicized problem. No policy differ-
ences on control are seen in California between the two parties or
their gubernatorial candidates. No one knows what the legislators
can or should do about the control of air pollution. Almost no
one is aware of any public agency involved in the control of air
pollution — much less what their activities are. No policy proposals or possible control steps could be specifically cited, except
that of smog devices for cars. No specific channels are viewed as
open either for complaints or personal action. In substance, most
people do not know how or even why they should become
participants in public policy decisions on air pollution control.[32]*

Policy stand as a concept draws attention to the policy preferences of
the public and, more generally, to the question of political support. In a
well-known definition, V.O. Key says that public opinion "may be taken to
mean those opinions held by private persons that governments find it prudent
to heed."[33] Politicians and specialists continually worry about what opinions
to heed in making policy decisions. Most policy makers in a democracy think
it desirable to make resource decisions in ways approved by the public. And,
more important, they are in fact presented by public opinion — though
uncertain, tentative, and complex in effect — with some policy directions,
limits, and sanctions.[34] The compelling question, however, is what public
opinions enter the air quality policy arena. Or, in other words, how do policy
makers know what the public endorses, prefers, or will support. The answer is
they don't, except in a gross, inexact, and perhaps erroneous manner.

There are now no reliable guides to the public's policy views on the
control of air pollution. Three ways, all suspect, provide some information.
Officials can proceed by hunch, intuition, and impression to estimate and
interpret public opinion. Yet these highly subjective views are open to error
and subject to argument. Or second, officials can depend on opinions
expressed by groups. Unfortunately, most people have no organized channels
in which to communicate their policy opinions. Citizen protest groups are
few in number, modest in organization, ill-informed on alternatives, and without the expertise or resources to counter special interests who can make their
positions known with vigor and persuasiveness. Citizens have few advantages in
clarifying and stressing their views as against the lobbying advantages of
polluters. And incentives for industry to lobby are high; as Benjamin Linsky,
former chief administrative officer for the Bay Area Air Pollution Control
District explained, "For every year an industry can defer spending $100,000
to install air pollution control equipment, they can usually keep $25,000 more
in their pocket before taxes."[35]

And Third, officials can consult the results of public opinion polls —
though I do not know of a single instance where a poll has been commissioned
to assist in the setting of air quality standards. Most newspaper and commercial polls on pollution do not probe deeply into public attitudes but rather

tabulate marginal distributions to superficial awareness or policy choice questions. Anyone familiar with the limitations of polling data has to be dissatisfied with the incomplete and partial character of these results. Either/or responses on a few questions, except in the most general sense, do not tell officials much about basic public preferences or willingness to support control measures.[36]

The public's interest and preferences for environmental quality and specifically the control of air pollution are policy stands that have not been effectively communicated to policy councils. The task of measurement is not, however, a simple one. The man-in-the-street does not have the information, time, expertise, in short, the ability to develop well thought-out, consistent, or detailed policy positions on how to control air pollution. Policy innovation must by necessity be the work of specialists. The public nonetheless can react to major policy programs, prospective or actual, and indicate whether they support the cost, method, and objective. The prevailing assumption of many politicians and specialists is that, of course the public wants clean air, BUT they are unwilling to pay for it.[37] This supposition needs to be closely studied, especially as applied to the indirect costs of controls for stationary sources. Moreover, in his first national survey on pollution, George Gallup concluded, "People are never eager to pay additional taxes, so these figures must be regarded as very encouraging. When these percentages are projected to the national adult population, it is clear that large potential funds are available to improve our environment."[38] To develop specific measures of support is beyond the task of this paper and depends further on the context and purpose of the survey. Nevertheless, a survey of personal responses should devote a number of questions to direction, intensity, and latency of support for air pollution control policies.

A Proposal for Regional Surveys

For setting — and perhaps enforcing — air quality standards, I propose that regional surveys of personal responses become one focus of deliberations on control policies. Surveys have demonstrated extraordinary utility as research instruments. Within the last two decades, the survey method has become the most important research procedure in the "behavioral sciences." I do not wish to present a "do-it-yourself" manual on the major techniques of survey research. Many excellent treatises are available on the requisite skills.[39] Instead, I recommend that surveys be contracted out to experienced polling firms and that these firms compete in open bidding for survey contracts.

While techniques of survey research can be readily applied, the survey design poses problems that are not technical in nature. Simply, the data from the survey can be no better than the questions put to the respondents. Survey objectives must be well thought-out; appropriate measures must be developed; useful questions must be written; and so forth. It is at this point that a team of consultants should draft general objectives, measures, and questions necessary to collect reliable and valid data on personal responses to air pollution. Questions on personal responses must proceed from what we want to know about the effects of air pollution and what we do know about social science measurement — especially questionnaire construction, scaling, index formation, and multivariate analysis.

Three objections can be raised to the proposal for regional pollution surveys. The first centers on the scientific credentials of survey research: that is, can survey results be accepted as objective data? A second contends the public cannot satisfactorily interpret or evaluate policy problems. And the third dismisses pollution surveys as expensive, irrelevant, even nonsensical exercises for the setting or air quality standards. The three objections merit at least a brief comment, for each represents a possible major drawback to regional pollution surveys.

First, the scientific credentials of survey research require but brief defense. Practitioners of opinion research have successfully demonstrated their claim to "scientific measurement of opinion." If we posit the major concerns of science as reliability and validity, several remarks follow. Reliability is judged by the reproducibility of a measurement, while validity has to do with whether the respondent's real opinion is discovered. Considerable evidence exists that survey results with a known margin of error can be reproduced and that polling measures correspond to the actual views of respondents. Perhaps the best known example can be found in the relation of pre-election polls and election statistics. Bernard Hennessy presents an excellent summary of these poll-elections results:

> The margin of error between the predicted popular vote and the actual vote for a candidate is a rough test of the validity of the prediction for that election. For a number of elections, these errors may be taken collectively as a gross estimation of poll validity. Similarly, the differences between or among poll forecasts for the same political race constitute a test of reliability. When aggregated, they may give us clues as to the general reliability of polls.
>
> The average error for all 245 national, sectional, state, and local election predictions made by the Gallup Poll from 1936 to 1950 was four percentage points. Since 1948, however, Gallup and the other major political pollsters have vastly improved their polling techniques and their understanding about political behavior. In six presidential and congressional elections, 1950 through 1960, the Gallup agency has averaged an error of less than 1 per cent. [40]

The second objection takes aim at the public's demonstrated lack of rationality, objectivity, and capacity to form consistent and well thought-out policy positions. Why should we poll prejudices and uncertain opinions toward the effects and control of air pollution? Of course, such surveys could provide a form of popular participation in setting air quality standards. Yet aside from the symbolic significance of such rituals, why should we rely on the views of the public? On pollution matters, the executive and administrative agencies are clearly the best qualified to see the problem as a whole and to prepare a coherent program as well as to handle the technical details of control. However, the public can express better than anyone else what it is they need and want. And herein lies the strength of public surveys:

> *The chief advantage of the polls is that, in an age of increasing strain upon traditional democratic procedures, they have made a constructive technical contribution by reflecting sensitively and flexibly the currents of public feelings, by making this information available to political leaders in a way which is neither rigid nor mandatory, and by testing the claims of special interests to represent the desires of the people as a whole. These are services performed by no other agency, and they should not be underestimated.*[41]

Opinion surveys serve to impress the public's concerns and priorities on the policy process.

As to doubts about the policy values of the surveys, any one of three possible contributions would warrant whatever expense and effort are involved. First, certain kinds of personal responses — as previewed earlier — could provide criteria for the formulation of air quality standards. Second, survey results would require that politicians and specialists review the impact of pollution on citizens in addition to evaluating the costs of control on "progress" and economic interests. There has been an eclipse of citizenship to the point that the public's priorities are seldom voiced and become in effect marginal to control decisions.[42] Survey results could provide a new calculus for the weighting of control goals, strategies, and actions. And third, pollution surveys should increase the pressure to adopt and enforce effective control programs, for they will register an increasing public concensus on the need for pollution control.[43] If counted, public concerns and preferences could provide major political incentives and sanctions in the setting of air quality standards. Moreover, there is no reason for those who draft air quality control programs to make impressionistic guesses on what the public wants or will support.

Notes

[1] Ido De Groot, "Trends in Public Attitudes Toward Air Pollution," *Journal of the Air Pollution Control Association* (October, 1967), 681.

[2] Jean Schueneman, "Air Pollution Control Administration," in *Air Pollution* (2nd Ed.; Vol. III), ed. Arthur Stern (1968), p. 721.

[3] For example, see discussions of air pollution politics by Edmund Faltermayer, *Redoing America* (1968), pp. 38-90, 205-227; and Howard Lewis, *With Every Breath You Take* (1965), pp. 222-263.

[4] See *Air Quality Criteria for Particulate Matter*, National Air Pollution Control Administration, U.S. Public Health Service, Washington, D.C. (January, 1969), Publication Number AP-49, and *Air Quality Criteria for Sulfur Oxides*, National Air Pollution Control Administration, U.S. Public Health Service, Washington, D.C. (January, 1969), Publication Number AP-50.

[5] "Managing the Environment," Report of the Subcommittee on Science, Research, and Development to the Committee on Science and Astronautics, U.S. House of Representatives (1968), p. 17.

[6] *Air Quality Criteria for Particulate Matter, op. cit.,* p. 7-13.

[7] Leslie Chambers, "Classification and Extent of Air Pollution Problems," in *Air Pollution* (2nd Ed.; Vol. I), ed. Arthur Stern (1968), pp. 17-19.

[8] *Ibid.,* p. 19.

[9] De Groot, *loc. cit.*

[10] *Restoring the Quality of Our Environment,* Report of the Environmental Pollution Panel, President's Science Advisory Committee, The White House (November, 1965), p. 43.

[11] Ronald Riker, "Strategies for Measuring the Cost of Air Pollution," in *The Economics of Air Pollution,* ed. Harold Wolozin (1966), pp. 91-92.

[12] John Middleton, "Public Policy and Air Pollution Control," Speech presented at the Penjerdel Regional Conference, Swarthmore College, Swarthmore, Pennsylvania, June 11, 1969, p. 3.

[13] Arie Jan Haagen-Smit, "Air Conservation and the Protection of Our Natural Resources," Proceedings of National Conference on Air Pollution, Public Health Service, Division of Air Pollution, Washington, D.C. (1963), Public Health Service Publication No. 1002, p. 173.

[14] Dorwin Cartwright, "Public Opinion Polls and Democratic Leadership," in *Public Opinion and Propaganda,* eds. Daniel Katz, Dorwin Cartwright, Samuel Eldersveld, and Alfred Lee (1964), p. 27.

[15] Robert Dahl, *Preface to Democratic Theory* (1956), p. 150.

[16] Jane Stein, "Priorities in Pollution: The SST and the Smogless Car," *The Washington Monthly* (February, 1969), p. 34.

[17] See Norton Long, "Power and Administration," and Herbert Simon, Donald Smithburg, and Victor Thompson, "The Struggle for Organizational Survival," in *Bureaucratic Power in National Politics,* ed. Francis Rourke (1965), pp. 14-22, 39-48.

[18] Murray Edelman, *The Symbolic Uses of Politics* (1964), pp. 23, 24-25, 26.

[19] *Ibid.,* p. 37.

[20] Theodore Lowi, *The End of Liberalism* (1969), p. 92.

[21] Jane Schusky, Lester Goldner, Seymour Mann, and William Loring, "Methodology for the Study of Public Attitudes Concerning Air Pollution," Paper presented at the 57th

Annual Meeting of the Air Pollution Control Association, Houston, Texas, June 21-25, 1964, p. 1.

[22] See John Pastier, "Conservation Group Wins Significant Legislative Battles," *Los Angeles Times,* Sunday, September 2, 1969, Section J, p. 15, col. 1.

[23] See Angus Campbell, Philip Converse, Warren Miller, and Donald Stokes, *The American Voter* (An Abridgement, 1964), pp. 97-108.

[24] W.W. Stalker and Charles Robison, "A Method for Using Air Pollution Measurements and Public Opinion to Establish Ambient Air Quality Standards," *Journal of the Air Pollution Control Association* (March, 1967), 143.

[25] Louis Harris, "New Priorities on Spending," Washington Post Co. (1967).

[26] Quoted by Robert Cahn, "Poll Finds Alarm Over Pollution," *The Christian Science Monitor,* Tuesday, March 11, 1969, p. 13, col. 1.

[27] James Beck and Marianna Stapel, "The Role of the Public in the Control of Air Pollution," in *Air Pollution Project: An Educational Experiment in Self-Directed Research, Summer 1968,* Associated Students of the California Institute of Technology, pp. 198, 195-22.

[28] W. Bryon Rumford, "The Politics of Pollution," *Journal of the Air Pollution Control Association* (July, 1966), 360.

[29] Stephen Ayres, "The Citizens Role in Air Pollution," Speech before the American Medical Association's National Congress on Environmental Health Management on April 26, 1967. Reprinted by U.S. Public Health Service, U.S. Printing Office: 1967 0-282-360, p. 15.

[30] See for example, Robert Lane, *Political Life* (1959), or Lester Milbrath, *Political Participation* (1965).

[31] Gilbert Seldes, "How Can We Get Action For Cleaner Air Through Public Communications?" Proceedings of National Conference on Air Pollution, Public Health Service, Division of Air Pollution, Washington, D.C. (1963), Public Health Service Publication No. 1002, p. 346.

[32] Ronald O. Loveridge, "Air Pollution and the Public Will," *California Air Environment* (April-June, 1969), 5.

[33] V.O. Key, *Public Opinion and American Democracy* (1961), p. 14.

[34] For example, Ray Kovitz has written, "After six years of Board activity, the MVPCB (Motor Vehicle Pollution Control Board) realizes that it can go no faster than public acceptance of its program." Ray Kovitz, "Gaining Public Acceptance for California's Auto Smog Control Program," *Journal of the Air Pollution Control Association* (January, 1967), 27.

[35] Quoted by Ernie Cox, "A Plea for Tougher Smog Laws," *Oakland Tribune,* Sunday, July 13, 1969, p. 25, col. 2.

[36] See Sidney Verba, Richard Brody, Edwin Parker, Norman Nie, Nelson Polsby, Paul Ekman, and Gordon Black, "Public Opinion and the War in Vietnam," *American Political Science Review* (June, 1967), 317-333.

[37] An illustration – Ray Kovitz declared, "Every Californian is against smog. He is more than willing to pass laws that will get rid of it. The problem arises when implementation of those laws costs him money." Kovitz, *op. cit.*, 26.

[38] Cahn, *loc. cit.*

[39] For a specific attempt to relate survey to public attitudes toward air pollution, see Schusky, *et al.*, "Methodology for the Study of Public Attitudes Concerning Air Pollution." Some of the most important general works on survey research include Hadley Cantril, *Gauging Public Opinion* (1944); Herbert Hyman, *Survey Design and Analysis* (1955); Leslie Kish, *Survey Sampling* (1965); and Frederick Stephan and Philip McCarthy, *Sampling Opinions: An Analysis of Survey Procedure* (1958). I would also recommend Herbert McClosky, *Political Inquiry* (1969) for an excellent discussion of the nature and uses of survey research in political science.

[40] Bernard Hennessy, *Public Opinion* (1965), pp. 89-90.

[41] John Ranney, "Do Polls Serve Democracy?," in *Public Opinion and Communication* (Enlarged Edition), eds. Bernard Berelson and Morris Janowitz (1953), p. 141.

[42] See Robert Pranger, *The Eclipse of Citizenship* (1968), especially pp. 39086.

[43] Most political observers agree that there is an increasing public consensus on the concept of clean air as a necessary requirement for improving the quality of our total environment. A recent editorial in a conservation magazine stated directly the position I find to represent the developing air pollution stance of the American people: ". . .we question why the richest nation on earth should be anything but publicly forthright and agressive about stopping the national disgrace of pollution NOW. The country should not wait until the impact on environment is worse, not until the problems have been studied to death, nor until the last buck has been made from causing the pollution itself." (See "Mr. Hickel Has A Choice," *Sierra Club Bulletin* (July, 1969), 2.) Yet despite public anguish and anger, the problem of air pollution – acccording to many sources – is getting worse. Pollution surveys should provide a political clout for new laws and new emphasis on pollution controls.

Comments on Paper Entitled "Types, Ranges, and
Methods for Classifying Human Behavioral Responses
to Air Pollution" By R. O. Loveridge, Ph.D.

By A. Steven Frankel, Ph.D.
Assistant Professor of Psychology
Department of Psychology
University of Southern California

As I see it, the theme of the paper is as follows: the individuals who are most affected by air pollution, and who may or may not be the most concerned about its effects, seem to be the least informed, the least considered by those who are in a position to do something about the problem, and the least powerful when it comes to taking action. Loveridge makes a case for survey research on this population to investigate personal, social, economic, and political effects of air pollution, so that hard data may be presented to those in power regarding these issues, in hopes of moving them to action.

I concur fully with Loveridge's position, and only wish that there were some way in which he could have been more emphatic about it, especially the part about powerlessness. I feel very strongly that the name of the game is, has been, and will continue to be power, especially economic power. Therefore, research in the economic sphere (e.g., will people vote down contracts from public agencies to "polluters"; does air pollution have a significant effect on business deals) will, in my opinion, prove to be most valuable.

My only criticisms of Loveridge's paper involve sins of omission rather than commission. I realize that more specific coverage of some issues, and the addition of others, would present time problems as regards symposia, but there may be some real benefit in presenting a few concrete research examples (e.g., how about following a random sample of newcomers to a highly polluted area, and comparing what happens to them to a matched sample of newcomers to a relatively non-polluted area? Some relevant dependent variables include familial and marital disharmony, financial change, incidence of "psychological disorder," incidence of physical disorder, and some measure of "psychological comfort.")

The other omission which I think needs to be mentioned is the need for controlled laboratory research on psychological or behavioral responses to air pollution. While survey research has its place, and often provides information which may be helpful, laboratory research may produce some more striking data. For example: If we could control pollution in a lab, we could study human interaction (peers, families, supervisor-supervisee, etc) as well as intra-human behavior (e.g., whether smog is a "stressor," whether it will have the same long term effects on human behavior that other "stressors" do). Finally, and perhaps most important in my opinion, is the necessity to differentiate between attitudes and overt behavior. Loveridge cites the polls of the body politic when suggesting that survey research can provide accurate data, but I am afraid that the issues we are faced with are not quite comparable to a national election. Thus, I agree fully that survey research is necessary, but I would argue strongly for controlled laboratory studies as well.

In conclusion, let me offer this outline regarding the goals of research in general in the area of air quality standards:

I. Exploration of the psychological (behavioral) effects of pollution.
   A. Intrapersonal ("psychological comfort," incidence of psychological disorders as well as physical disorders).
   B. Interpersonal and social (including recreation, family interaction, peer interaction, job interaction).
   C. Economic
   D. Political (can knowledge of the effects of pollution be used to get people to be willing to actually give up something to clean up the environment, such as voting for an "anti-smog" candidate).
II. Exploration of the means of communicating the results of the research carried out under I. (above) so as to provide optimal motivation to the population.
III. Exploration of the means of communicating the results carried out under I. (above) so as to provide optimal motivation for legislators to take action.

Again, I agree with Loveridge's position and I look forward to the symposium and to further discussion of this topic.

TYPES, RANGES, AND METHODS FOR CLASSIFYING HUMAN PATHOPHYSIOLOGIC CHANGES AND RESPONSES TO AIR POLLUTION

By Ian T. T. Higgins, M.D.*
and
James R. McCarroll, M.D.†

Although deleterious effects on human health are most often cited as the rationale for air pollution control, clearly demonstrable responses to usual levels of pollution are difficult to demonstrate in "normal" urban residents at usually-encountered pollution levels. Evidence for such ill effects, ranging from possible to incontrovertible, comes from a variety of types of observations. Some of these, in turn, suggest possible mechanisms by which certain pollutants, or combinations of pollutants, may be exerting effects upon human health and function. Three types of evidence have been offered in support of the thesis that human physiologic mechanisms may be altered by commonly-encountered atmospheric pollutants. All this evidence, regardless of source, is, in varying degree, indirect for one reason or another.

One type of evidence offered to support a claim for health effects depends on the response of lower animals to various substances to which they may be exposed — alone, or in varying combinations. Such studies, of course, offer tremendous advantages to the investigator, in terms of his ability to carry out types of observations which would be impossible with human beings. Since many of the basic physiologic processes of the living cell are shared by all forms of life, such observations unquestionably contribute to our understanding of basic physiologic processes. The flexibility permissible in such experimental approaches offers the possibility of studying types of mechanisms and responses which could not be similarly tested in man. Animals may be exposed to toxic doses of substances far exceeding those permissible in human experimentation. Animals may be sacrificed to study structural alterations produced by the experimental process. Organs, or sections of organs, may also be examined in various stages of activity leading to advances in knowledge which may have equal relevance for man. Through this approach, we may learn a great deal about the effects of specific pollutants (or combinations) on the various protective mechanisms of the respiratory tract, such as

*Professor of Epidemiology and Community Health Services, The University of Michigan School of Public Health, Ann Arbor, Michigan 48104.

† Professor of Preventive Medicine and Director, Division of Environmental Health, Department of Preventive Medicine, University of Washington School of Medicine, Seattle, Washington 98105.

ciliary movement, mucous secretion and movement, mechanism of action of pulmonary macrophages, and the role of various enzyme systems in detoxifying, or otherwise affecting, pollutants.

In spite of these very real advantages, it is obvious, however, that we are on relatively shaky ground if we cite this type of evidence as a basis for insisting on air pollution control as a menace to human health. The wide — sometimes astoundingly wide — difference in response of different species of animals to idential compounds makes extrapolation from responses of lower animals to man extremely dangerous. Although a guinea pig may be killed, or seriously incapacitated, by exposure to a given level of an air pollutant, it is a weak argument if we confront an industry demanding it spend many thousands or millions of dollars to remove this compound from its effluent, based upon the supposed effect on human health. The structure, function, reaction, and exposure of many of our commonly-used laboratory animals may vary widely from the reactions of Homo sapiens. As an example, penicillin — probably our single most useful antibiotic in controlling human disease — although almost completely benign to man, toxicologically is highly toxic to the guinea pig, which can be immediately killed by doses which we regularly administer to our human patients. Without belittling the very real contribution of animal studies to physiolgic knowledge, we must admit that DISEASE IN MAN must be our major criterion in this problem.

The second type of evidence offered to support air pollution as a hazard to human health obviates many of these difficulties by dealing directly with man. These studies derive from the exposure of human volunteers — or, occasionally, populations industrially or otherwise accidentially exposed to high concentrations of various pollutants. In these cases, it has frequently been possible to demonstrate various types of adverse effects to pollutants, ranging from death in a few accidental exposures to a variety of physiologic changes, depending on the pollutant used. These physiologic changes may involve many different mechanisms. Irritation of exposed mucous membrane surfaces, such as the eye, the nose, and the upper respiratory tract, can be produced by a variety of substances among the various groups of oxidant compounds present in photochemical smog. Other physiologic changes, such as increased airway constriction, which can be measured by appropriate instrumentation, can be demonstrated by other common pollutants, such as oxides of sulfur. Other objective measurements of the effect of certain pollutants can be demonstrated on the central nervous system, such as the effects of carbon monoxide on a variety of cerebrally controlled functions.

The major disadvantage of this type of study is that, almost invariably, they require doses of pollutants far in excess of those usually encountered in normal urban air to produce a physiologic response. To return to our original question, as to whether these substances do, in fact, produce disease in man at the levels at which we encounter them, we must look at man in his natural environment — in other words, epidemiologic studies. Although the authors of this paper consider themselves epidemiologists and, therefore, prejudiced, we nevertheless feel that the principle evidence relating customary levels of urban air pollution to human disease derives from epidemiologic studies. These, of course, are observations of actual human populations, in their usual environments, carrying out the activities they would normally pursue. If such observations do, in fact, indicate a relationship between disease in these populations, and variations in levels of air pollution, such data would be the best available on which to make a judgment on air quality standards.

We recognize, of course, the defects inherent in epidemiologic studies — the difficulty, or impossibility, of selecting control populations, the inability to control numberous variables, and the difficulties in selecting truly representative samples. Nevertheless, we strongly feel the evidence derived from this type of observation provides the strongest and most incontrovertible evidence for an effect of air pollution on disease in man. We therefore propose to review the evidence derived from studies of this type as that most relevent to our present task of developing air quality standards.

The Acute Episodes

The most convincing evidence of an effect of polluted air on health, and certainly the earliest to be recognized, comes from the acute episodes of pollution. There have been a number of these, several of which have become famous: The Meuse Valley, Belgium, in 1930; Donora, Pennsylvania, in 1948; and London, England, in 1952. It is less widely known that London has suffered from severe killing fogs at least since the nineteenth century. An appendix in the Ministry of Health's report of the London fog in 1952 lists the more important episodes which have occurred since 1873. Since 1952, London has experienced lethal fogs in most years until 1963. The characteristics of all these episodes have been elevated levels of both smoke and sulfur dioxide and probably elevated levels of sulfur trioxide, sulphuric acid, sulphates, carbon monoxide, and carbon dioxide, which have not been routinely measured. In each fog there has been increased mortality and morbidity. The increase in mortality is for all causes, but is particularly marked for certain specific causes of death, notably bronchitis, pneumonia, other respiratory diseases and coronary heart disease. Both sexes and all ages were affected in the 1952 fog, but in other episodes those over 65 seem to have been more often affected (Scott, *et al.,* 1964). Low temperature has sometimes been considered to be a more important cause of mortality than pollution (Russell, 1926). But in comparing the two effects in the 1962-1963 episode, Scott concluded that, for every increase of 100 micrograms/cubic meter of smoke, there were seven additional deaths, while for every fall of one degree temperature there were four additional deaths.

Morbidity has been measured in several ways. The most useful has been claims for sickness benefit received by the Ministry of Pensions and National Insurance. These apply to the great majority of the working population for illness which results in four, or more, days of absence from work. Other indices which have been used have been hospital admissions statistics, applications for admission to hospital by means of the Emergency Bed Service Bureau, notifications of pneumonia, and observations of interested general practitioners on the number of new cases seen during an episode, compared to other times. All of these indices are subject to a number of obvious criticisms on the grounds of accuracy or completeness.

Accepting that acute episodes of pollution may have effects on mortality and morbidity, the question remains to what extent are the aged, the young, and the ill being predominantly affected? Can pollution initiate respiratory disease, or does it merely exacerbate and administer a coup de grace? There is insufficient evidence to enable one to answer these questions. It would seem clear, however, that the effect is predominantly one of exacerbation. Evidence for an initiating effect can be obtained from the fact that cattle were affected in the Meuse Valley and in London. In Donora, 10 dogs, 3 cats, and 2 canaries were said to have died.

Episodes of pollution similar to those which have occurred in London have occurred in New York City. These, too, have been associated with an excess number of deaths and an excess morbidity. In general, the excess mortality seems to have been smaller than that reported for London. Thus, there were reported 200 excess deaths in 1953, 800 in 1963, and 168 in 1966. Analysis by age suggests that death has been mainly confined to the 45-64 and 65-and-over age groups. The excess deaths in New York City have been attributed to influenza, pneumonia, vascular lesions of the nervous system, cardiac disease, and "all other" causes of death, but not to accident, homicide, suicide or deaths in early infancy. Attempts to measure morbidity have been made by analysis of clinic visits to major city hospitals. It has been possible to show an increase in the number of emergency visits for bronchitis and asthma on some occasions, but not on others. From the point of view of research planning, the acute episodes should presumably be looked on as "natural experiments", which are decreasingly likely in the future. As such, every advantage should be taken to study them, should the opportunity arise. But an accident of excessive pollution can hardly be considered as a suitable topic for planned research. We therefore turn to the much harder problem of the effects of lower levels of pollution.

Day-to-Day Measurement of Mortality and Morbidity in Relation to Pollution

In order to define a lower level of pollution above which effects on health might be expected, a study was started in London by the Ministry of Health in 1958/59. Daily measurements of smoke and sulfur dioxide are related to daily deviations of death and illness from a moving average of 7 or 14 days. Correlations between the deviations and levels of pollution are then calculated for all causes of death, or sickness, and for selected causes. In this way, Martin and Bradley (1960) and Martin (1964) have been able to show a significant positive correlation between black suspended matter in the atmosphere and the daily number of deaths; a slightly less, but still significant, association between atmospheric SO_2 and deaths; and a significant negative association between visibility and deaths. Significant correlations have also been demonstrable between pollutant concentrations and indices of morbidity.

An analysis of daily mortality in relation to daily pollutant levels in this country has been made in California by Mills (1960), who analyzed daily deaths from cardiac and respiratory disease in 1956, 1957, and 1958, in relation to daily oxidant and temperature in Los Angeles. He considered daily deviations in the number of deaths from the monthly average and concluded that there were 400 excess deaths a year attributable to smog in Los Angeles. He noted that, in the temperature range 46°F to 104°F, ozone levels were correlated with temperature. It is known that high temperatures are associated with excess deaths. But Mills considered that heat deaths were unlikely to occur except at temperatures above 96°F. He therefore excluded days when the temperature was in this range from his analysis. It would appear to be debatable to what degree the deaths attributable to smog were really due to lower temperatures than those excluded by Mills.

Hechter and Goldsmith (1961) analyzed a number of environmental factors and daily cardiac and respiratory deaths in Los Angeles County for the same three years — 1956-1958. They noted a distinct seasonal pattern, with some irregular oscillations. They removed the seasonal component by the technique of harmonic analysis and found that, after adjusting for auto correlation in the remaining residuals, air pollution, as measured by oxidant and carbon monoxide concentration, exerted no detectable influence on day-to-day mortality. It seems likely that the excess smog mortality found by Mills was really the result of an inadequately detailed analysis, and that the best evidence indicates that, however objectionable it may be, oxidizing pollution has no significant effect on mortality.

Studies of Hospital, Clinic or Disabled Patients

A number of studies of hospital or clinic patients with varying degrees of disability have been carried out to assess their reactions to changes in air pollutant levels. The earliest studies were those of Lawther and his colleagues (Waller and Lawther, 1955 and 1957). Daily diaries were kept by a group of patients attending the bronchitis and emphysema clinic at St. Bartholomew's Hospital in London. Each patient entered a daily record of whether his chest felt better than usual, worse than usual, or the same as usual. These indices of well-being were correlated with daily measurements of smoke and SO_2 made at St. Bartholomew's Hospital. A close correlation was found between the concentration of pollutants and the clinical condition of bronchitis during the winter months, but there appeared to be no correlation during the summer when the pollution fell to lower levels.

77

More objective methods of assesment of patients by tests of lung function have been carried out in this country by Spicer and his colleagues in Baltimore (Spicer, *et al.*, 1962). These workers studied a group of 150 patients aged 20-65 with chronic obstructive airways disease who lived within a defined area of Baltimore. They have presented information on intensive studies of small numbers of patients drawn from this total group. In the winter of 1960-1961, they first studied seven patients five days a week, Monday through Friday, for 14 weeks. Various tests of ventilatory capacity and airways resistance were carried out, and clinical and sputum examinations were made. They then studied 14 patients daily for 47 consecutive days in a similar manner. Whether these 14 patients included any of the previous patients is not clear. Daily changes in lung function, sputum, or well-being were related to daily changes in particulates, SO_2, and nitrates. The authors' general conclusion was that the patients became better or worse together, and they considered that this suggested that they were affected by some common factor in their environment. It was not, however, possible to identify any one pollutant as being the factor involved.

Similar studies of patients in Cincinnati conducted by Shepherd and his colleagues (Carey, *et al.*, 1958; Shepherd, *et al.*, 1960) yielded similar results. These workers carried out lung function studies three times weekly on 10 cardiorespiratory cripples for 10 weeks from October, 1956, and related the findings to pollution and environmental measurements. There was an immediate response to suspended particulate matter, which was consistent with broncho-constriction, but the most significant features were transient depression of pulmonary pressures and carbon monoxide uptake.

The Long-Term Implications of Air Pollution Episodes

An attempt was made to assess the effect of the Donora episode of 1948 on subsequent mortality and morbidity over the next 10 years by Ciocco and Thompson (1961). Essentially, a review of the probability sample of one-third of the town, which had been studied in 1948 by the Public Health Service, was carried out. The study group was divided into two groups on the basis of answers to the question, "Were you affected by the smog of October 28-31?" During the next 10 years, those who had answered "yes" in 1948 to this question experienced higher mortality rates than those who had answered "no". But most of this excess appears to have been experienced by those with chronic illness before the fog. The authors state that the differences in mortality between those who were acutely ill and those who were not are sharply reduced, if the comparison is limited to the persons reporting no prior chronic illness in 1948. Among those with no prior chronic illness, the differences in mortality rates between those who were affected and those who were not were, however, consistently higher in men up to age 64 and in women at all ages. The authors also comment that the prevalence in the residents of the survivors of the 1948 Donora survey in 1957 was no different from that of two neighboring communities and from the current residents of Donora. In view of the excess mortality over the past 10 years, one might have expected a lower prevalence in the survivors, and this equality may well reflect an increased frequency of disease.

Geographical Comparisons of the Effects of Pollution on Health

1. International Comparisons

The remarkable differences between mortality for bronchitis in different countries has given rise to the hypothesis that they are, in part, the result of differences in air pollution. That there are differences in mortality from chronic respiratory disease between countries, for example, between the United Kingdom, United States, and Scandinavian countries, which are not merely the result of diagnostic practices, seems clear (Christensen and Wood, 1958; Mork, 1962; Reid, *et al.,* 1964). The use of the term "bronchitis", however, greatly exaggerates differences between the United States and United Kingdom, and a more realistic comparison can probably be obtained by comparing the mortality for all respiratory diseases. This reduces a forty-fold to a three- or four-fold difference in each sex.

Differences in mortality have been supported by differences in morbidity and differences in lung function in a number of comparative surveys (Reid, *et al.,* 1964; Holland, *et al.,* 1965; Mork, 1961; Olsen and Gilson, 1960; Ferris and Anderson, 1964). The tendency has been to ascribe any differences between different countries in symptom prevalence of lung function values which are found, which cannot be explained by differences in smoking or differences in height, to an urban factor, or to air pollution. But such a conclusion is often debatable. Instrumental and observer variation, selection into or out of a particular job, ethnic differences, or differences in intelligence may all confound a simple attribution of a difference between countries to air pollution.

2. Regional Comparisons Within Countries

There is a well-marked trend of increasing mortality rates for chronic lung disease death rates from rural through urban to city areas (Reid, 1964). This is so both for males and for females, but is more marked for males. In the United States there is a similar, though far less impressive, trend for males, but none for females (Manos, 1957). These ovservations have stimulated research into air pollution and death rates.

The earliest types of study of regional differences used readily available data. Consequently, the extent of the study depended on the degree of coverage of the country for pollution measurements. These were initially related to mortality rates in different areas.

Pemberton and Goldberg (1954) correlated bronchitis death rates with indices of air pollution in 35 county boroughs of England and Wales for 1950, 1951, and 1952. There was a significant correlation of mean annual sulfur dioxide concentration (lead peroxide candle) and bronchitis in men in each of these three years, but no significant correlation in women. There were no significant correlations between "total solids" and bronchitis mortality for either sex.

More indirect indices of air pollution based on fuel comsumption in defined built-up areas were used by Daly (1954 and 1959). Daly correlated death rates for six causes in 1948-1954 of men aged 45-64 in the 83 county boroughs of England and Wales. There was a relatively close association between air pollution and bronchitis and a relatively low association between pollution and lung cancer mortality. High correlations of bronchitis mortality were also found with social class, over-crowding, population density, and education. But using partial correlation, Daly showed that, when these were allowed for, there still remained a highly significant correlation between pollution and bronchitis. As a generalization, Daly concluded that, "since two-thirds of the variation in bronchitis death rates could be attributed to the combined effects of air pollution and social factors (as measured), it follows that about one-third of the variation in mortality could be attributed to air pollution and and one-third to social factors." Smoking was also included in this study, the results from a survey of a quota sample carried out by Research Services, Ltd., being used. The results were, however, analyzed only in relation to lung cancer and not in relation to bronchitis mortality.

Stocks (1959) studied mortality from bronchitis, lung, stomach, intestinal, and breast cancer in relation to population density, atmospheric deposit, and smoke in 53 county boroughs of England and Wales and certain administrative areas in Yorkshire and Lancashire. He found that bronchitis and lung cancer were highly correlated with both types of pollution in the county boroughs. In the Yorkshire and Lancashire areas bronchitis was correlated with deposit but not with smoke, while the reverse was the case for lung cancer. Lung cancer in Lancashire areas was also strongly correlated with population density. Stomach cancer was significantly related to deposit and smoke. This he attributed to pollution of food by dirty air. Neither intestinal nor breast cancer appeared to be significantly related to air pollution indices.

A Nationwide Study of Sickness Absence
in Relation to Pollution

The Ministry of Pensions and National Insurance Enquiry into the Incidence of Incapacity for work included a section on the Relationship of Incapacity and Air Pollution. In this enquiry, incidence of incapacity for work in different areas and occupations in 5 per cent of all employed men and 2½ per cent of all employed women in England and Wales and Scotland were studied. Incapacity rates for bronchitis, influenza, arthritis and rheumatism, and psychoses and psychoneuroses were related to levels of smoke and sulphur dioxide pollution (Clifton 1964). The findings for bronchitis were analysed for the Greater London conurbation and for the rest of Great Britain separately. There was a sugnificant correlation between bronchitis incapacity in men aged 35-54 and the average levels of smoke (suspended matter) and sulphur dioxide in high-density residential districts. South Wales had far more bronchitis incapacity than could be accounted for by the air pollution found there. In the Greater London conurbation, which was studied in more detail and where air pollution levels were more closely related to socio-economic factors, there was a significant correlation between bronchitis incapacity and both forms of pollution for all age groups taken together and for men aged 35-44 and 55-59. There was also in both studies more incapacity from arthritis and rheumatism in heavily smoke-polluted areas. Influenza incapacity was greater in those areas with higher pollution levels over Britain as a whole, but this did not apply within the Greater London conurbation, nor was there any assiciation between pollution and psychoses and psychoneuroses.

The interpretation of this important study is difficult. The clear correlation shown between bronchitis incidence and smoke and SO_2 suggests the importance of pollution. But the fact that the relationship also held for arthritis and rheumatism suggests that some factor other than pollution must also be playing a role. The obvious factor which seems to have been inadequately controlled is social class.

Studies of Uniform Industrial Groups in Different Places

In the United States, Holland and Stone (1964) carried out an investigation of Bell Telephone System employees in three East Coast cities and compared them with a similar group studied by Deane, *et al.*, 1965. In a previous paper they compared the findings with comparable groups studied in England (Holland, *et al.*, 1965). The Forced Expiratory Volume ($FEV_{1.0}$) standardized for age and smoking was consistently higher in the United States groups than in the British, and this is in the direction one would expect if air pollution was having an effect on lung function.

Dohan (1960 and 1961) studied respiratory diseases in female employees of RCA plants in five cities for which air pollution data were available. Using insurance records he was able to obtain incidence rates in hourly employees for respiratory illnesses which lasted for 7 days and over the years 1955, 1957, and 1958. He observed a high correlation between the average rates for the 3 years of such illnesses and the mean level of "suspended particulate sulphates". Age distribution, conditions of work, social and climatic factors did not appear to be responsible for the five-fold variation in incidence of these illnesses between the cities.

Comparison of Two Towns with Different Amounts of Pollution

A comparison of respiratory disease and lung function in two towns in Pennsylvania with contrasting levels of air pollution was carried out by Prindle and his colleagues in 1959 and 1960 (Prindle, *et al.,* 1963). The village of Seward with relatively heavy air pollution was compared with the village of New Florence with relatively light air pollution. Information on respiratory disease in both towns was collected by questionnaire, chest X rays, and lung function tests, and these indices were correlated with pollutant measures. Average dust fall in Seward was 3.2 times and SO_2 6.2 times that of New Florence. Age and height adjusted values of most lung function tests were remarkably similar for both sexes in both towns. The only difference was in the average airways resistance and in airways resistance times volume, both of which were higher in Seward than in New Florence. This difference could be due to the difference in air pollution between the two towns; but there were other differences between the two populations than that of air pollution concentrations. Seward contained a larger proportion of coal miners than New Florence and the chest radiographs taken in the survey indicated that the prevalence of pneumoconiosis was twice as high in the more polluted town. Either mining or pneumoconiosis could have accounted for the small difference in airways resistance. Nor were smoking habits considered. One would like to have seen a comparison confined to non-miners and women which also allowed for smoking habits.

Differences in Disease Mortality and Morbidity Within Towns

Reid (1956 and 1958; Fairbairn and Reid 1958) have drawn attention to the existence of mortality and morbidity differences in different areas of a single large city. They correlated these differences with an index of pollution based on fog frequency. As already described, the Enquiry into the Incidence of Incapacity carried out by the Ministry of Health extended these observations. Attempts have also been made in this country to relate differences in pollutant concentrations within large cities to mortality and morbidity.

In Buffalo, Winkelstein and his colleagues carried out a survey of respiratory disease in relation to air pollution (Winklestein, 1962). Suspended particulates, settleable solids, and oxides of sulphur in the ambient air were monitored by a network of air-sampling stations during a two-year period, July, 1961, to June, 1963. Mortality for the years 1959-1961 for persons aged 50 years and over for all causes and for bronchitis, asthma, or emphysema mentioned anywhere on the death certificate were related to concentrations of suspended particulates (Winklestein, *et al.,* 1967). In a subsequent paper, total and respiratory mortality rates were related to sulphur oxides (Winklestein, *et al.,* 1968). The results were presented for persons aged 50-69 and 70 and over by socio-economic level, this being based on median family income of the census tract of residence. Four levels of pollution and five levels of economic status were used. There was a positive association between suspended particulate level and total mortality in both men and women aged 50 and over, though in men aged 70 and over the relationship was much attenuated. There was a significant correlation between suspended particulate level and mortality from respiratory disease in men aged 50-69, but the number of deaths at age 70 and over in men was too few for analysis. Presumably, the number of respiratory deaths in women were insufficient for analysis at all ages. The authors also found a strong inverse relationship between socio-economic status and positive relationship between particulate concentration and deaths from lung cancer. In their subsequent paper dealing with sulphur oxides, the authors found a positive association between sulphation and mortality for chronic respiratory disease in men aged 50 and over within the two lowest economic levels. Total mortality from all causes, however, showed essentially no association with sulphation level. Nor did these workers find any support for the hypothesis that there is a synergistic effect of high suspended particulates and sulphation.

The main deficiency of this important study is that smoking habits and occupation were not considered. It is unfortunate that these were not studied in a sample of decendents in order to determine to what extent they might have confounded the conclusions. To date, no one within this country has apparently tried to obtain information on such important factors in studies of mortality by interviewing the next of kin of the decendent. Such an investigation would be worthwhile.

Winkelstein's study also provides some suggestive information about desirable lower levels of pollution which might be aimed at. Thus, mortality for respiratory disease was appreciably greater in his two higher particulate pollution classes. This would put the desirable average annual concentration of suspended particulates at 100 micrograms/cubic meter or below.

Studies of Air Pollution in Nashville, Tennessee

A more extended study of the effects of air pollution within a single city was carried out by the Public Health Service in Nashville, Tennessee (Zeidberg et al., 1964, 1967). An air pollution sampling network of 123 stations located throughout Nashville was set up. Dust fall, soiling, sulphation, and SO_2 were monitored and mortality and morbidity were related to levels of pollution.

All deaths occurring between 1949 and 1960 were studied. Total mortality, respiratory disease mortality (excluding tuberculosis and neoplasms of the respiratory system), mortality from influenza and pneumonia, lung and bronchial cancer, bronchitis and emphysema, and tuberculosis were related to pollution and socio-economic levels. Age-specific respiratory disease mortality from 25-74 years was directly related to the degree of exposure to sulphation. Age-specific mortality for all respiratory diseases including tuberculosis and respiratory neoplasms was directly related to the degree of exposure to sulphation. At all levels of sulphation, respiratory disease mortality was higher in men than in women, and this was especially marked for respiratory cancer.

With the exception of respiratory cancer, respiratory mortality was higher in non-whites at high and moderate levels of sulphation. Mortality for non-malignant respiratory disease was inversely related to socio-economic class when the degree of exposure to pollution was kept constant. Within the middle socio-economic class, non-malignant respiratory disease mortality was directly related to the degree of exposure to sulphation and soiling. But bronchial cancer mortality was inversely related and bronchitis and emphysema were not significantly related to the level of sulphation. In a subsequent paper on cardiovascular disease mortality, a consistent pattern of higher rates for all cardio-vascular disease was found with higher pollution (soiling) in women. The association of cardiovascular disease mortality and soiling was much less consistent in men, being noted mainly for "other myocardial degeneration" and to a lesser degree for hypertensive heart disease. The generally expected differences in cardiovascular disease with sex, race, and socio-economic status were noted. Mortality for cancer of the stomach, esophagus, and prostate appeared to be directly associated with the degree of exposure to suspended particulates in men. In women a similar association was observed for bladder cancer.

No observations were made on smoking habits of the decendents, nor was any study of mortality in relation to occupation made. As in all studies of this kind, the degree to which the stated pollution levels actually represented those experienced by the decendents during their lives is somewhat uncertain. Some attention to this might be given in any future studies by obtaining information on residence during the decedents lifetime from his next of kin.

Studies of Air Pollution in Genoa, Italy

One of the most interesting studies of the effects of pollution in different parts of a town has been conducted by Petrilli and his colleagues in Genoa, Italy. Genoa is a rapidly growing industrial city situated between the Appennines and the Mediterranean. As a result of its topography, the local climate may differ strikingly in different parts of the city. From 1954-1964 pollution has been assessed at 19 sites. Sulphation, SO_2, CO, lead, deposited matter, suspended matter estimated gravimetrically and by particle size and 3-4 benzpyrene have been monitored. Mean values for available data in 1962-64 showed an increase of about 20 per cent from 1954-1961. Respiratory symptoms were assessed using the British Medical Research Council's (M.R.C.) Questionnaire in women aged 65 years and over who were non-smokers and who had not worked in industry but who had resided for a long period in the same area. The results were also analyzed by floor level of residence. Strong associations were shown between the frequency of respiratory disease and the concentration of SO_2. Unfortunately, the analysis of this material is inadequate to form a firm judgment about the effectiveness with which the authors have standardized for socio-economic factors. They also present annual morbidity indices for bronchitis and other respiratory disease

(1961-1962) for seven districts where air pollution was measured continuously. The rates are in general related to the average annual SO_2 concentrations.

Comparison of Berlin, New Hampshire, with Chilliwack, British Columbia

Ferris and Anderson compared respiratory disease and lung function in representative samples of the inhabitants of two towns which differed in their degree of air pollution. In 1961 they studied a 1:10 sample of adults living in Berlin, New Hampshire, and two years later, in 1963, they carried out a comparable study of a 1:7 sample of the inhabitants aged 25-74 of Chilliwack, British Columbia. In each survey the methods were comparable and the observers the same. One major cause of differences between surveys was therefore largely eliminated. Information about respiratory symptoms and smoking habits was obtained by using a prototype M.R.C. respiratory symptoms questionnaire and the forced expiratory volume and peak respiratory flow were used to assess ventilatory lung function. On the basis of the age-specific prevalence of symptoms in both surveys, expected prevalences for non-smokers were calculated. While from the multiple regression equations on age and height for the $FEV_{1.0}$ and Peak Flow Rates for Berlin, expected lung function values for men and women were similarly calculated for Chilliwack. The investogators found that there was little difference in respiratory disease prevalence for male non-smokers from that expected, but in women the prevalence was slightly below expectation. Lung function values were also consistently slightly higher in men and women in Chilliwack for all smoking categories. These differences are in a direction one would expect if pollution was exerting an effect. But as the authors point out, they might well be due to other differences between the populations; notably, differences in ethnic background.

Unfortunately, only two measures of pollution, namely, dust fall and sulphation rates, were used. The results for the latter were expressed as equivalent SO_2 levels in parts per million. Even in Berlin the SO_2 pollution appears to be low, a monthly variation of from 0.02-0.03 parts per million (or approximately 60-90 micrograms/cubic meter) being recorded for a nine-month period, August, 1960, to April, 1961. Dust fall, however, would indicate that Berlin is a more polluted town than the sulphur dioxide levels suggest. Over the same months dustfall varried around 45 tons/square mile/30 days with a peak up to 75 tons/square mile/30 days in February, 1961, which indicates that in this respect Berlin is a moderate to heavily polluted town. In 1967 a follow-up study of the Berlin population was carried out. All those seen in the original study were reviewed and a new cross section of the town was sampled in order to investigate cohort changes. During the course of this survey, measurements of particulates were made with hi-volume samplers. It should be possible, therefore, to obtain a better idea of the levels of particulate pollution to which the inhabitants are subjected when these data are analyzed.

Longitudinal Observations of Various Groups in Relation to Air Pollution

A number of studies have followed groups of early bronchitics to assess the effect of changes in air pollution on the progress of the disease.

Since 1961 Fletcher has been making regular observations on some 1,000 men aged 30-59 working at the London Transport Engineering Works at Chiswick and Acton and at the Post Office Savings Bank in Hammersmith, London. Measurements of ventilatory capacity and sputum volume have been made every 6 months since 1961 and on each occasion the men have been asked about any chest illnesses. Many mentioned "bronchitis", "chest cold", "catarrh". In order to study the nature of these illnesses and to see what changes in sputum characteristics or ventilatory capacity accompanied them, a closer surveillance of 85 men out of the whole group was carried out during the winter of 1962/63. Illnesses were classified on a symptomatic basis into coryza, chest colds, muco-purulent bronchitis, wheezy attacks, "flu" or acute respiratory disease. The incidence of these illnesses was related to both smoke and sulphur dioxide in the atmosphere in which the men lived. There was no significant relationship with low temperature after standardizing for smoke and SO_2. The prevalence of illness was more closely related to smoke than to SO_2. The concentration of smoke over the period was in the range of 200-400 micrograms/cubic meter daily with peaks up to 1100 and 800 micrograms/cubic meter. SO_2 concentration was between 400 and 500 micrograms/cubic meter with peaks to 2947, 1200, 900 and 600.

Bates and his colleagues have studied a group of 216 Canadian Veterans in Halifax, Montreal, Toronto, and Winnepeg. Observations have been made on respiratory symptoms and lung function, and lung function testing was repeated at monthly intervals. The cities differ in climate and air pollution. Halifax, Montreal and Toronto are relatively dirty cities, but Winnepeg, while it has a good deal of dust fall from prairie dust, has no industrial or domestic pollution. The group of veterans in Winnepeg was similar in age, overseas service record, occupation, age of starting cigarette smoking, average cigarette consumption and age at onset of cough to those from Toronto, Montreal and Halifax. But they had a lower incidence of hemoptysis, less severe dyspnoea, fewer chest illnesses and better preservation of ventilatory capacity and diffusing capacity than the other men. Whether these differences were due to differences in air pollution between the cities is debatable. As the authors point out, they might have been due to climatic variation, differences in the natural history of chest infection, a higher standard of medical treatment and supervision at one center compared with another, or presumably to initial differences in the type of veteran coming under observation.

Studies of Children

A number of studies of the effects of air pollution have been carried out in children. Children offer the two salient advantages that they are not engaged in occupations which expose them to dusts, fumes or gases and at least under 10-12 years of age the majority of them do not smoke cigarettes.

Wahdan (1962) studied children in Sheffield, England, at that time a fairly polluted city, and the Vale of Glamorgan, Wales, an area of unpolluted countryside. He found that the frequency of sinus opacity and otitis media was higher in children in Sheffield. It seems possible that the differences might have been due at least in part to differences in socio-economic level between the children in the two areas. Toyama (1964) and Watanabe (1965) studied school children in Kawasaki and Osaka, two polluted cities in Japan. In each study the results of simple lung function tests in children attending schools in polluted areas were compared with those obtained in children attending school in less or minimally polluted areas. Lower peak flow rates were attributed in each case to higher pollution exposures. No differences were found in vital capacity between the schools.

A National Survey of Health and Development was carried out in Britain between 1946 and 1961 (Douglas and Waller 1966). All births which occurred during the first week of March, 1946, were used to obtain a sample of children who were followed until they reached school-leaving age in 1961. The relationship of respiratory infections to residence in areas of high or low pollution was analyzed in 3866 of the children. The main object of the investigation as far as pollution was concerned was to assess the importance of prolonged exposure to polluted air at lower levels than those which occur in exceptional episodes. Pollution was determined for separately identifiable towns in terms of coal consumption per unit area. Subsequently, the classification was validated by measurements of smoke and sulphur dioxide. Four levels of pollution were used: very low, low, medium and high. The addresses of the children were recorded on eight occasions between 1946 and 1961. On this basis it was possible to classify all the children in terms of their probable lifetime exposure to air pollution. Upper and lower respiratory illness and hospital admissions from these or any other causes were recorded when the children were aged 2 and 4 by health visitors and during examinations by school doctors when they were 6, 7, 11 and 15 years old. Causes of absence from school for more than one week were also noted. The results were simple and consistent: upper respiratory tract infections were not related to pollution; lower respiratory tract infections were. The frequency and severity of such infections increased with the amount of pollution. Boys and girls were similarly affected and, rather surprisingly, no difference was found between children in middle- and working-class families. The association between pollution and lower respiratory tract infection was found at each age examined, and the results of the school doctors' chest examinations suggested that it persisted at least until school-leaving age.

This important paper can be criticized on the grounds that the assessments of levels of air pollution were made in the year ending May, 1952, when the children were 5 to 6 years old, whereas illnesses were recorded up to age 15. Furthermore, the assessment of pollution was based on the indirect measure of fuel consumption per unit area. Inevitably, such a measure of pollution will be correlated with density of population. This in turn is correlated with susceptibility to infection and consequently it is uncertain how much the associations with pollution which were found are really due to associated liability to infections. The study is, however, remarkable example of the value of information collected for one purpose being used for another. It would seem desirable that studies of a similar nature, but deliberately aimed at studying the effect of air pollution, should be conducted in this country.

Douglas and Waller give probable levels for smoke and SO_2 levels experienced by the children in their four pollution classes. The lowest level was around 70 micrograms/cubic meter for smoke and 90 micrograms/cubic meter for SO_2. Higher rates of illness were noted in all the higher pollution classes. Consequently, if one accepts these figures as accurate measures of pollutant levels and the suggestion that the higher rates of illness in the higher pollution classes are due to pollution, one is forced to conclude that the desirable level of these two constituents is under an average of 100 micrograms/cubic meter per annum.

An interesting piece of supporting evidence for the importance of home residence in early life on the liability to respiratory disease was obtained by Rosenbaum (1961) who correlated the incidence of respiratory disease in National Servicemen with their home localities before call-up. He found that an industrial home background had an adverse effect on subsequent illness during National Service. The most likely explanation appeared to be a lower resistance in those from certain localities to the acquisition of infection on leave. The recruit also may have brought more infection with him since the highest incidence was in recruits.

Lunn and his colleagues (1967) studied patterns of respiratory illness in infant school children in Sheffield, England, and related these to pollution levels in the areas in which the children lived. Eight hundred and nineteen children aged 5 to 6 years attending certain selected schools were studied. Information on previous illness and present symptoms were collected from the parents, an examination by a physician and the forced expiratory volume and forced vital capacity were measured. Upper and lower respiratory tract infections were associated with area which generally reflected the pollution levels. Illness was less common in clean than in dirty areas. In contrast to the area findings, social class, number of children in the house, and sharing bedrooms appeared to have little influence on respiratory illness. The $FEV_{0.75}$ and FVC were unaffected by socio-economic factors or area except in the most heavily polluted district where there was a significant reduction. A past history of persistent or frequent cough, colds going to the chest, pneumonia and bronchitis was associated in all areas with a reduction in FEV and a weaker association occurred with FVC in all areas combined. In addition to children aged 5 to 6 years, Lunn and his colleagues also studied a group of children aged 10 to 11 years, but they have not yet published their findings on these older children. Further similar studies would seem to be strongly indicated of children living in areas of contrasting pollution in this country.

More recently, Holland and his colleagues (1969) have studied some 10,000 school children in four areas in Kent. Peak expiratory flow was found to be related in area of residence, social class, family size and a past history of pneumonia, bronchitis or asthma. These four factors appeared to act independently and the effects were additive. The findings suggested that environment early in life can produce adverse changes which may persist and contribute to the development of chronic respiratory disease.

Miscellaneous Studies of Air Pollution Effects

Migrants

The suggestion that air pollution plays a role in the development of lung cancer has been made as a result of observations on immigrants into New Zealand, South Africa and Israel. Eastcott (1956) analyzed deathrates for British immigrants into New Zealand and compared them with deathrates for native-born New Zealanders (other than Maoris). The populations were those as defined at the 1951 Census and the deaths for 1949-1953. The rates were standardized for place of residence since immigrants tend to live in the towns. Standardized rates for nine sites showed no appreciable differences, but the rates for lung cancer were higher in immigrants. They were, moreover, higher in those who immigrated after the age of 30 than in those who immigrated earlier in life. The conclusion drawn from this study was that the British environment must have had a special effect on the production of lung cancer. Total tobacco consumption has been practically the same in both countries since 1900 and air pollution associated with urbanization seemed the most likely explanation of the findings. However, total tobacco consumption has been found to correlate poorly with lung cancer in several countries and cigarette consumption may have differed more in the two countries than the total tobacco consumption suggests.

In South Africa similar studies of immigrants were conducted by Dean. Mortality rates for lung cancer in British immigrants were compared with those for Union-born men for the years 1947-1956. At ages 45-64 years the mortality among male British immigrants was found to be higher than that of Union-born men. In each area studied the total mortality was 44 percent higher than that expected from the age-specific deathrates for Union-born men] in the relevant residential district. In contrast, there was little difference between the observed and expected mortality among British immigrants at ages 65 and over, or among other immigrants in either age group. The mortality of male British immigrants was slightly lower than that recorded of the same period for England and Wales. The total South African mortality was substantially lower than the British mortality. The differences cannot be explained by differences in cigarette smoking since the South Africans have been among the heaviest consumers of cigarettes in the world. Neither the type of tobacco consumed nor the somewhat longer cigarette butt discarded in South Africa appear to explain the difference which, as in the case of New Zealand, has been atrributed to the relatively higher exposure to pollution of migrants in Britain before they emigrated.

Studies of migrants into the United States are at present being carried out by Haenszel and others and should throw further light on the importance of early-life residence in relation to mortality from lung cancer and chronic respiratory disease. One difficulty encountered in such studies should, however, be noted. It is fairly well established that migrants differ in a number of respects from those who remain at home. The assumption made in such studies that the emigrants would have had mortality rates typical of their country of origin had they not emigrated is almost certainly false. It is clearly important to define accurately the characteristics of migrants more precisely than has usually been done in retrospective studies of mortality.

Twins

Cederlof has investigated the importance of constitutional and environmental factors in the pathogenesis of chronic respiratory disease in twins in Sweden. Information on zygosity, respiratory symptoms, smoking habits and residential history was obtained by mailed questionnaires. Information was complete on 9230 twin pairs of whom 1553 couples were discordant with regard to residential history but at the same time concordant with regard to smoking. In the whole population, urban/rural morbidity ratios for bronchitis were above unity for men and women smokers and non-smokers. In the group which was concordant for smoking but discordant for residence, morbidity ratios for bronchitis were above unity in smokers but below unity in non-smokers both for monozygotic and for dizygotic twins. The authors interpret these findings to mean that a specific urban factor is in some way interacting with smoking. While the study does not indicate the factor, air pollution was suspected. Studies of twins in relation to air pollution are at present being conducted in this country and should prove valuable. It would also be desirable to carry out similar studies in the United Kingdom where pollution levels are so much higher then in Sweden.

Exercise Performance

The effect of oxidant air pollution on athletic performance was studied in Los Angeles by Wehyrle and his colleagues in 21 competitive high school cross-country races from 1959-1964. Running times tend to improve throughout the season. Consequently, team performance was assessed by measuring the percentage of boys who failed to improve their running times when compared with the previous meet on the same course. Carbon monoxide, temperature and humidity were not related to performance. But oxidant level in the hour before the race was correlated with team performance, the higher the level, the lower the performance. Somewhat lower correlations were also obtained with suspended particulate levels one hour before the race.

In a further study of athletic performance and oxidant levels carried out in Seattle, the performance of 115 men who ran a weekly two-mile run over a two month period was analyzed in relation to pollution (Koonitz 1968). No direct correlations were found with total oxidants, particulates, temperature of humidity. Performance did, however, appear to be affected by change in either oxidant level or temperature. A rise in temperature of 20°F or of oxidants of 3-4 parts per million from the previous week was associated with significantly lower performance than on stable days. It was not possible to separate the effect of pollution from that of temperature. The authors suggest that the apparent lack of any correlation with absolute pollution level compared to the effects which have been noted in the previous study may have been due to the fact that pollution is relatively low in Seattle.

The authors believe that the aforementioned epidemiologic studies, in spite of the numerous shortcoming and deficiencies we have cited, nevertheless constitute the best available evidence for an effect of urban air pollution on human health. Despite the wide variety of study designs, certain general conclusions on the effect of air pollution on human health can be reasonably inferred from these data:

1. There is a wide variation in susceptibility of different persons to some air pollutants.
2. Persons with certain types of pre-existing or underlying disease conditions — particularly cardiac and respiratory conditions — are often more susceptible to the effects of air pollutants than well persons. The additional stress imposed by some types of air pollution may aggravate the pre-existing disease, even to the point of precipitating death. Various functions of persons without underlying structural disease may also be affected under some conditions by increased levels of air pollution. This may occur at all ages, may be related primarily to pulmonary function, or may perhaps involve other organ systems.
3. Air pollution may be an additive factor aggravating the effect of other substances such as tobacco smoke, asbestos, etc., in initiating disease.
4. Evidence for initiation of disease in well persons, without other demonstrable toxic exposures, is much less clear. Nevertheless, the total thrust of most of the studies cited above indicates that, under some conditions, some types of air pollution can actually initiate structural and persistent disease in some persons.
5. The evidence cited above would also clearly suggest that different types of air pollution may produce quite different physiologic responses.

Our conclusions must, therefore, be that there is no one pathophysiologic change which can be attributed to air pollution. The effect of air pollution on human function is insidious and protean, and may be manifested in many different ways. It is also impossible to specify exact ranges which may be called "safe", because of the wide variation in susceptibility, both of normal people and of persons with pre-existing disease conditions.

We must conclude that it is impossible to signify a lower limit for any single pollutant which will be "safe" for all persons. The total thrust of the evidence cited would also suggest that no one pollutant acting alone can be held responsible for the observed changes in health, and that some type of synergistic or additive mechanism is probably present.

It is obvious that our most important need at the present time is for more information on all of these points. Particularly needed are measurements of both morbidity and mortality, which should be made now for use in the future, as denominators so that the effects of the Clean Air Act can be measured.

Acknowledgements

Much of the material in this paper is from a background document prepared by Dr. I.T.T. Higgins for the Task Force on Research Planning in Environmental Health Science for the National Institute of Environmental Health Sciences. The report of the Task Force will be published by the Government Printing Office. The original material for this Background Document, as well as others prepared for that Report, will be deposited in the National Library of Medicine which will supply copies on request. Supported by P.H.S. Contract 86-68-142 from the Health Services and Mental Health Administration.

References

General Reviews of Effects of Air Pollution

1. Air Quality Criteria for Particulates, Dept. H.E.W. P.H.S. Consumer Protection & Environmental Health Service. National Air Pollution Control Administration, Arlington, Virginia, 1969.

2. Air Quality Criteria for Sulphur Oxides. Dept. H.E.W. P.H.S. Consumer and Environmental Health Service. National Air Pollution Control Administration, Arlington, Virginia, 1969.

3. Anderson, D.O.: The effects of air contamination on health. Part I. Canad. Med. Assoc. J. 97:528-536, September 1967.

4. Anderson, D.O.: The effects of air contamination on health. A review. Part II. Canad. Med. Assoc. J. 97:585-593, September 1967.

5. Anderson, D.O.: The effects of air contamination on health. Part III. Canad. Med. Assoc. J. 97:802-806, September 1967.

6. Ferris, B.G., Jr., and Whittenberger, J.L.: Environmental hazards. Effects of community air pollution on prevalence of respiratory disease. New Eng. J. Med. 275:1413-9, December 1966.

7. Goldsmith, J.R.: Effects of air pollution on human health. In: Air Pollution. Vol. I. Stern, A.C. (ed.), 2nd Edition. Academic Press, 1968.

8. Heimann, H.: Status of air pollution health research, 1966. Arch. Environ. Health 14:488-503, March 1967.

References referred to in the text:

1. Anderson, D.O., Ferris, B.G., and Zickmantel, R.: Levels of air pollution and respiratory disease in Berlin, New Hampshire, Amer. Rev. Resp. Dis. 90: 877-887, December 1964.

2. Angel, J.H., Fletcher, C.M., Hill, I.D., and Tinker, C.M.: Respiratory illness in factory and office workers. A study of minor respiratory illnesses in relation to changes in ventilatory capacity, sputum characteristics, and atmospheric pollution. Brit. J. Dis. Chest 59: 66-80, April 1965.

3. Bates, D.V.: Air pollution and chronic bronchitis. Arch. Environ. Health 14: 220-224, 1967.

4. Carey, G.C.R., Phair, J.J., Shephard, R.J., and Thomson, M.L.: The effects of air pollution on human health. Amer. Ind. Hyg. Assoc. J. 19:363, 1958.

5. Cederlof, R.: Urban factor and prevalence of respiratory symptoms and "angina pectoris." A study on 9,168 twin pairs with the aid of mailed questionnaires. Arch. Environ. Health 13: 743-8, December 1966.

6. Christensen, O.W. and Wood, C.H.: Bronchitis mortality rates in England, Wales and Denmark. Brit. Med. J. 1: 620-622, March 1958.

7. Ciocco, A. and Thompson, D.J.: A follow-up of Donora ten years after: Methodology and findings. Amer. J. Pub. Health 51: 155-164, 1961.

8. Clifton, M.: The national survey of air pollution. Proc. Roy. Soc. Med. 57: 1013-1015, October 1964.

9. Daly, C.: Air pollution and bronchitis. Brit. Med. J. 2: 687-688, September 1954.

10. Daly, C.: Air pollution and causes of death. Brit. J. Prev. Soc. Med. 13: 14-27, January 1959.

11. Dean, G.: Lung cancer among white South Africans. Brit. Med. J. 2: 852-857, 1959.

12. Deane, M., Goldsmith, J.R. and Tuma, D.: Respiratory conditions in outside workers. Arch. Environ. Health 10: 323-331, 1965.

13. Dohan, F.C.: Air pollutants and incidence of respiratory disease. Arch. Environ. Health 3: 387-394, 1961.

14. Dohan, F.C. and Taylor, E.W.: Air pollution and respiratory disease, a preliminary report. Amer. J. Med. Sci. 240: 337-339, September 1960.

15. Douglas, J.W.B. and Waller, R.E.: Air pollution and respiratory infection in children. Brit. J. Prev. Soc. Med. 20: 1-8, 1966.

16. Eastcott, D.F.: The epidemiology of lung cancer in New Zealand. Lancet 1: 37-39, 1956.

17. Fairbairn, A.S., and Reid, D.D.: Air pollution and other local factors in respiratory disease. Brit. J. Prev. Soc. Med. 12: 94-103, April 1958.

18. Ferris, B.G. and Anderson, D.O.: Epidemiological studies related to air pollution: A comparison of Berlin, New Hampshire, and Chilliwack, British Columbia. Proc. Roy. Soc. Med. 57: 979-983, October 1964.

19. Hechter, H.H. and Goldsmith, J.R.: Air Pollution and daily mortality. Amer. J. Med. Sci. 241: 581-588, 1961.

20. Holland, W.W., Halil, T., Bennett, A.E., Elliott, A. Factors influencing the onset of chronic respiratory disease. Brit. Med. J., 2: 205, 1969.

21. Holland, W.W., Reid, D.D., Seltser, R., and Stone, R.W.: Respiratory disease in England and the United States. Studies of comparative prevalence. Arch. Environ. Health 10: 338-345, February 1965.

22. Holland, W.W. and Stone, R.W.: Respiratory disorders in United States East coast telephone men. Amer. J. Epidemiol. 82: 92-101, 1965.

23. Koonitz, C.H.: Oxidant air pollution and athletic performance. In press, 1968.

24. Lunn, J.E., Knowelden, J., and Handyside, A.J.: Patterns of respiratory illness in Sheffield infant school children. Brit. J. Prev. Soc. Med. 21: 7-16, 1967.

25. Manos, N. U.S. Public Health Service Publication No. 562, 1957.

26. Martin, A.E. and Bradley, W.: Mortality, fog, and atmospheric pollution. Monthly Bull. Ministry of Health 19: 56-69, 1960.

27. Martin, A.E.: Mortality and morbidity statistics and air pollution. Proc. Roy. Soc. Med. 57: 969-975, 1964.

28. Mills, C.A.: Respiratory and cardiac deaths in Los Angeles smogs during 1956, 1957, and 1958. Amer. J. Med. Sci. 239: 307-315, 1960.

29. Ministry of Pensions and National Insurance. Report of an Enquiry into the Incidence of Incapacity for Work. Chapter 6: The relationship between incapacity and air pollution. Her Majesty's Stationery Office, London, 1965.

30. Mork, T.: A comparative study of respiratory disease in England and Wales and Norway. Norwegian Monographs on Medical Science. Norwegian Universities Press, 1962.

31. Olsen, H.C. and Gilson, J.C.: Respiratory symptoms, bronchitis and ventilatory capacity in men. An Anglo-Danish comparison, with special reference to differences in smoking habits. Brit. Med. J. 1: 450-456, 1960.

32. Pemberton, J. and Goldberg, C.: Air pollution and bronchitis. Brit. Med. J. 2: 567-570, September 1954.

33. Petrilli, F.L., Agnese, G., and Kanitz, S.: Epidemiology studies of air pollution effects in Genoa, Italy. Arch. Environ. Health 12: 733-740, June 1966.

34. Prindle, R.A., Wright, G.W., McCaldin, R.O., Marcus, S.C., Lloyd, T.C., and Bye, W.E.: Comparison of pulmonary function and other parameters in two communities with widely different air pollution levels. Amer. J. Pub. Health 53: 200-218, February 1963.

35. Reid, D.D.: Air pollution as a cause of chronic bronchitis. Proc. Roy. Soc. Med. 57: 965-68, 1964.

36. Reid, D.D., Anderson, D.O., Ferris, B.G., and Fletcher, C.M.: An Anglo-American comparison of the prevalence of bronchitis. Brit. Med. J. 2: 1487-1491, December 1964.

37. Reid, D.D.: General epidemiology of chronic bronchitis. Proc. Roy. Soc. Med. 49: 767-71, 1956.

38. Rosenbaum, S.: Home localities of national servicemen with respiratory disease. Brit. J. Prev. Soc. Med. 15: 61-67, 1961.

39. Russell, W.T.: The relative influence of fog and low temperature on the mortality from respiratory disease. Lancet 2: 1128, 1926.

40. Scott, J.A., Taylor, I., Gore, A.T., and Shaddick, C.W.: Mortality in London in the winter of 1962-1963. Med. Officer (London) 111: 327-330, June 1964.

41. Shephard, R.J., Turner, M.E., Carey, G.C.R., and Phair, J.J.: Correlation of pulmonary function and domestic microenvironment. J. Appl. Physiol. 15: 70-76, January 1960.

42. Spicer, W.S., Jr., Storey, P.B., Morgan, W.K.C., Kerr, H.P., and Standiford, N.E.: Variation in respiratory function in selected patients and its relation to air pollution. Amer. Rev. Resp. Dis. 86: 705-712, November 1962.

43. Wahdan, M.M. Ph.D. Thesis, University of London, 1962.

44. Waller, R.E. and Lawther, P.J.: Some observations on London fog. Brit. Med. J. 2: 1356-58, December 1955.

45. Waller, R.E. and Lawther, P.J.: Further observations on London fog. Brit. Med. J. 2: 1473-75, December 1957.

46. Watanabe, H.: Air pollution and its health effects in Osaka. Presented at the 58th Annual Meeting of Air Pollution Control Association. Toronto, Canada, June 20-24, 1965.

47. Wayen, W.S., Wehrle, P.F., and Carroll, R.E.: Oxidant air pollution and athletic performance. J.A.M.A. 199: 901, 1967.

48. Winkelstein, W.: The Erie County air pollution-respiratory function study. J. Air Pollution Control Assoc. 12: 221-222, May 1962.

49. Winkelstein, W.: The relationship of air pollution and economic status to total mortality and selected respiratory system mortality in man. Arch. Environ. Health 14: 162-169, January 1967.

50. Winkelstein, W., Kantor, S., Davis, E.W., Maneri, C.S., and Mosher, W.E.: The relationship of air pollution and economic status to total mortality and selected respiratory system mortality in men. II. Oxides of sulfur. Arch. Environ. Health 16: 401-405, March 1968.

51. Zeidberg, L.D., Prindle, R.A., and Landau, E.: The Nashville air pollution study. III. Morbidity in relation to air pollution. Amer. J. Pub. Health 54: 85-97, January 1964.

52. Zeidberg, L.D., Horton, R.J.M., Landau, E., Hagstrom, R.M., Sprague, H.A.: The Nashville air pollution study. V. Mortality from diseases of the respiratory system in relation to air pollution. Arch. Environ. Health 15: 214-224, 1967.

Discussion of Paper by Doctors Higgins and McCarroll
"Types, Ranges, and Methods for Classifying Human
Pathophysiologic Changes and Responses
to Air Pollution

By O.J. Balchum, M.D.

The knowledge of the respiratory effects of air pollution from epidemiological studies was critically analyzed by Doctors Higgins and McCarroll. The problems in evaluation of this evidence were reflected in phrases such as "difficult to demonstrate" and "ranging from possible to incontrovertible."

Studies of the effects on animals as well as occupational and accidental exposures in man have also yielded an important background of information. In setting the early air quality standards in California, very little more than such information was available.

The measuring sticks of most epidemiologic studies have been mortality, respiratory symptoms, and lung function measurements. These are not disease processes, but the reactions of people to various stimuli. The authors point out that in case of acute air pollution episodes, the predominant effect is one of exacerbation of a previous disease process resulting in either death, or increase in respiratory symptoms or altered lung function.

There may be in the near future evidence of more specific reactions, if not disease processes, due to air pollution. Recent new information on the reduced lysozyme content of tears resulting from eye irritation due to pollutants, the IgA deficiency which may prove to be associated with chronic bronchitis, and the increased lipid peroxidation taking place in the lungs and red blood cells of animals exposed to NO_2 and ozone, respectively, may signify the identification of more specific indicators of response to air pollutants. An increased lipid peroxidation is possibly related to the lung deterioration from NO_2 inhalation, possibly even emphysema.

Such investigations are difficult to carry out in the sense that many associated factors cannot be controlled and there is a great deal of "background noise". People with asthma, for example, go or do not go to clinics or hospitals for a variety of reasons. A clearly defined group that goes specifically because of worsening of symptoms has never been delineated, isolated or defined in any study, to permit relating measurements to air pollution levels.

Perhaps, a clearer relationship between air pollution and lung reactions is being demonstrated in the studies in children in both Japan and England. This is important in the sense that degenerative lung conditions and smoking are probably almost totally absent initially in this population. The results tend to confirm conclusions from adult population studies.

The authors conclusions from epidemiologic evidence in adults are clearly stated, namely that exposure to air pollution results in symptoms of lung irritation and alterations in function, and aggravates the symptoms of previous disease, and even tips the balance toward death. It is evident that individuals vary widely in susceptibility and there are important differences in response due to sex, socio-economic level and many other factors. The type of physiologic alterations are related to the nature of the air pollutant and type of mixture. No clear lower or safe level has been demonstrated. Epidemologic studies of humans have not as yet permitted the conclusion that air pollutants initiate structural and persistent disease in some persons. Such urban air pollution effects have not been demonstrated in man. However, it must be strongly suspected that such effects might possibly be occurring after decades of living in urban polluted air. Most of us would agree that this is a possibility after considering the evidence from recent animal studies involving NO_2 and ozone.

One of the most frequent causes of death is by accident. It might be possible to design an epidemiologic study of the lung tissue alterations from living many years in polluted air, by study of the inflated lungs from accident victims. Even some post-mortem functional evaluation could be carried out. The decendents of such subjects could be questioned and the evidence from various medical examinations during the lifetime of the subjects might be helpful. Even spirometric data is becoming available on several thousands of people in some cities. The relationships of sex, aging, smoking, and other factors could be explored. The nature of the lung changes in the various entities of the emphysema-bronchitis complex might be defined and related to urban and industrial air pollution exposures. Chemical analysis of the lung tissue and blood could be carried out.

Secondly, if occupational medicine could be conducted in a way to make possible long-term, systematic observations of the results of occupational exposures a great deal of information on the effects of specific pollutants would be available.

Now that the genetic influence of alpha-1 antitrypsin deficiency has been detected in emphysema, another factor has been added to this multifactor problem. A frontal approach is indicated and would keep a National Lung Institute busy for several decades. It is doubtful whether this dream will become reality and probably we will have to be satisfied in the future with rough estimates as to whether air pollution actually causes diseases of the lungs.

Comments on Paper Entitled "Types, Ranges, and Methods for Classifying Human Pathophysiologic Changes and Responses to Air Pollution," by Doctors Higgins and McCarroll

By Jack D. Hackney

This extensive review of epidemiologic evidence that human physiologic mechanism may be altered by commonly encountered atmospheric pollutants gave many examples of the tremendous difficulties and handicaps inherent in the large-scale epidemiologic approach to the problem. These deficiencies were pointed out by the authors repeatedly. In the studies reviewed, it appeared to be especially difficult to obtain an adequately matched population which would provide control for socioeconomic classes, smoking history, kind of pollution, and uniformity of exposure (even within the same city), permanence of inhabitance, and occupation. Without adequate control, valid conclusions appear to be questionable.

Even the clear-cut examples of effects of excessive concentrations of pollution on cardiopulmonary patients, such as occurred in Donora (1948), or London (1952), are of limited value. For future similar studies, the further careful demonstration on a large scale that ill patients are made worse by stress (including irritating atmosphere) is expensive confirmation of what is now common knowledge and is not likely to lead to further advances in ways to manage air pollution effects. The use of gross indicators, such as mortality or hospitalization rates, probably greatly reduce the sensitivity of such studies; effects on subjects with less severe disease or on hypersensitive normals may be totally overlooked.

Certainly, efforts to refine and utilize these techniques need to be continued, but because of the difficulties, I favor in addition an alternative approach which uses less gross indicators and emphasizes detailed data and smaller scale studies. Smaller studies can better utilize sensitive physiologic measurements on representative population samples exposed to (1) single or multiple specified pollutants, or (2) ambient polluted atmosphere. Admittedly, if the wrong tests are chosen or if appropriate sensitive indicators are not available, even this type of approach will not give adequate information. For example, exposing subjects to a challenge of one cigarette would not affect the measured vital capacity (maximum volume of air that can be inhaled or exhaled) or the forced expiratory volume at one second (the volume of air that can be forcibly exhaled at one second). However, blood carbon monoxide would definitely increase, body surface temperature would be affected, and increase in airway resistance could be demonstrated in some participants.

Regarding the exposure of human volunteers to known pollutants, the authors state, "The major disadvantage of this type of study is that, almost invariably, they require doses of pollutants far in excess of those usually encountered in normal (sic) urban air to produce a physiologic response." I ask, *why* is this? It might be because: (1) we have not used or found the right test or indicator, or (2) if only mean values of findings in experimental and control groups are compared, some of the subjects who actually respond may be overlooked.

Hopefully, the right tests can be found and measurements with sensitive indicators of physiologic function on representative samples of normal and diseased populations challenged with known pollutants accomplished in the near future. This will offer the advantage of "before" and "after" studies on the same subjects and minimize the risk that hyper-reactors will be overlooked because they were lost in the population averages (if only the means of different groups were considered). In such studies special attention is needed to determine whether the time course of the body's response to pollutants is persistent or transient. Persistent responses would be especially suspect as initiators or exacerbators of disease.

It seems prudent to give some thought to what sensitive indicators are now available, what information they have supplied, and the problem of what is still needed. In the Los Angeles area all of us have experienced obvious eye and upper airway effects at the height of pollution. This is because we are very sensitive and excellent detectors and respond as transducers. Also, trained observers (pulmonary physiologists, athletes, coaches), as well as the untrained (patients with pulmonary disease) have experienced obvious interference with respiratory function, especially noticeable with exercise. Certainly, these observations are subjective and open to misinterpretation. Still, considering the usual objectivity and specialized training of many of these observers, I feel that this evidence cannot be dismissed immediately as only "testimonials." Perhaps further collection, documentation, and study of this type of evidence by questionnaire is needed, but a more appealing direction is to search for sensitive objective indicators. In this direction of research, I believe several recent developments offer promise of success at both the organ and cellular levels.

The airways offer resistance to air flow. With ordinary methods the resistance measured is mostly that of the large airways. Just beyond and in the smaller (peripheral) airways, the business of gas exchange is transacted and it is in this area that minimal disease may have maximum consequences. However, significantly increased airway resistance in this latter area is relatively undetectable. Recent findings in man and animals suggest that changes in peripheral airway resistance can be detected by measuring the pressure required to expand the lungs at different breathing frequencies. At high frequencies, inhomogeneities in resistances between the parallel units is exaggerated and more pressure is required to move the same volume.

The lungs are the meeting place of fresh air and blood. Even slight alterations in function tend to result in mismatching of the air and blood. A sensitive index (measurement of nitrogen dissolved in body fluids) is available for assessing this dysfunction, and with recent innovations in the methods of measurement, this index can be more widely implemented in human and animal studies.

What are the effects of toxic levels of oxidants on living organisms? As an example of a relevant animal model, rabbits and mice usually die within three to four days from continuous exposure to 100% oxygen (at ambient pressure). The lungs show extensive damage. At the cellular level, recent work suggests that long before structural changes are seen, DNA synthesis in lung alveolar cells is greatly decreased in this high oxygen exposure. I believe that these studies serve as a useful model for predicting that similar effects will result from exposure to other oxidants such as ozone or oxides of nitrogen. While the technique for labeling DNA synthesis is not suitable for human studies, it is a sensitive tool for further animal uses.

We need to continue trying to find (and validate) appropriate sensitive indicators of pollution effects even if they are only suitable for studies on limited population samples of man or animals. Some of these also might become applicable to large epidemiological type studies.

Because of the population and automobile explosions, it is entirely possible that we will be unable to control pollution of the general atmosphere. It is certain that we cannot regulate the cigarette smoker's personal atmosphere. So, even if epidemiological studies establish an association between existing pollution and human disease, and even if air quality standards ultimately are agreed on and set, we will still need to understand the way (the mechanism of action) in which pollutant substances alter physiology. This knowledge can suggest therapeutic experiments (or clues) to help reverse or prevent these physiologic changes. Unfortunately, large scale epidemiological studies alone probably never can provide this kind of detailed knowledge.

ABATEMENT STRATEGY AND
AIR QUALITY STANDARDS

By Robert E. Kohn*
Assistant Professor of Economics
Southern Illinois University, Edwardsville

Air quality standards should be based, in part, upon abatement capability. It is appropriate therefore that a symposium on standard setting examine the relationship between air quality and the strategies for achieving it.

Three strategies for air pollution control may be considered. The first is the imposition of controls on those pollution sources from which the necessary reduction in emissions can be achieved at the least total cost of control. Whether the burden of control is subsidized or borne by emitters is of secondary importance. The more important goal is the attainment of a set of air quality standards with the minimum expenditure of resources.

A second strategy, the levying of pollutant fees on emitters, would tend to cause relative prices to reflect both the cost of abatement plus a charge for uncontrolled emissions. This would reduce pollution both by inducing voluntary abatement by emitters and through a shift in comsumption toward goods involving less pollution in their production or use. The third strategy involves control methods with non-market payoffs in addition to air quality. For example, a mass transit system would not only reduce air pollution but highway congestion as well. The third strategy is a total systems approach incorporating standards for many parameters of environmental quality.

In this paper I will emphasize the first strategy, that of determining how air quality standards can be achieved at minimum cost. The progression from this strategy to a fee system, specifically intended to alter relative costs between goods, will be illustrated. Only brief mention will be made of the total systems approach.

The first strategy lends itself to model building. Three basic variables are involved: I. A vector of air quality standards, q; II. A set of abatement activity levels, x; III. Total cost of control, z.

Three mathematical programming models incorporating these variables will be presented in this paper. The first is a cost effectiveness model in which q is given and x and z are determined. The second is a budget constraint model

---

*The writer's doctoral research on air pollution was supported by U.S.P.H.S. Air Pollution Fellowship Number F3 AP 37, 270-01, 2. Computer time was made available under N.S.F. Grant Number G-22296 to Washington University. Post doctoral research has been supported by funds and release time from the Business Division, S.I.U. – Edwardsville.

in which z is given and q and x are determined. In the final model, q, x, and z are simultaneously determined. The second and third models are in contrast to the view, sometimes implied, that air quality standards should be set independently of abatement capability. (see 8)

Cost Effectiveness Model

This model will be illustrated, using data on emissions and control methods for the St. Louis airshed in 1975. The target of air quality standards, expressed in annual averages, is the vector

$$1)\ \bar{q}\ =\ \begin{bmatrix} 75.\ \mu g/m^3 \text{, suspended particulates} \\ .02 \text{ ppm, sulfur dioxide} \\ 3.1 \text{ ppm, total hydrocarbons} \\ .069 \text{ ppm, oxides of nitrogen} \\ 5.\ \text{ppm, carbon monoxide} \end{bmatrix}$$

The standards for particulates and sulfur dioxide are those adopted by the Missouri Air Conservation Commission (22, p. 1). The standards for hydrocarbons and nitrogen oxides are the levels that prevailed at the St. Louis CAMP station in 1964, while the standard for carbon monoxide represents an arbitrary 20% improvement over the 1964 level (21, p. 6).

The cost effectiveness model, in matrix notation is

$$2)\ \text{Minimize } z\ =\ cx$$
$$3)\ \text{Subject to} \quad Ux \quad = s$$
$$4) \qquad\qquad\quad Ex\ -\ a\ =\ 0$$
$$5) \qquad\qquad\qquad\quad Ma\ \leqslant q\ -\ b$$
$$6) \qquad\qquad\qquad x \qquad \geqslant 0$$

The variables are z, the total cost of abatement, and x, the set of control method activity levels. The latter will be illustrated for one of the sources in the St. Louis model, identified as Source 2, the combustion of number two diesel fuel (6, p. 64). For this source, two abatement or control activity variables are considered, the first of which is a continuation of the existing state:[1]

$$7)\ x_2 \quad = \text{ gallons of number two diesel fuel burned in buses.}$$
$$8)\ x_{2A} \ = \text{ gallons of number two diesel fuel replaced by number one diesel fuel in buses.}$$

---

[1] Two alternative but equivalent models were suggested in the writer's dissertation (6, pp. 27-30). This paper presents what I called the Bergstrom model, with the illustrative equations and coefficients transformed where necessary.

The cost of these control methods are

9) c_2 = $ .00 per gallon.

(For simplicity, the existing control methods, regardless of whether they represent some voluntary type of control or non-control, are assumed to have zero unit cost.)

10) c_{2A} $ .015 per gallon

Equation 2 in the matrix model above is the summation of all "N" possible control method activity levels times their unit costs. The sum, which represents total cost of abatement, is the quantity which is minimized subject to the remaining constraints.

It is estimated that the magnitude of Source 2 in 1975 will be 7,250,000 gallons of number two diesel fuel (6, p. 64). Since source magnitudes are assumed fixed, the total of the activity levels for control methods 2 and 2A is accordingly constrained as follows:

11) $x_2 + x_{2A}$ = 7,250,000 gallons

Correspondingly, equation 3 in the matrix model constrains control method activity levels for each source to the estimated source magnitude. The set of L source magnitudes is denoted by the (L x l) vector, s. The capital letter, U, is an (L x N) matrix, whose element $U_{m,n}$ is unity when control method "n" applies to source "m" and zero otherwise. Equation 3 has the effect of requiring some level of control for each source, even if it is only a continuation of the existing state.

The emissions from Source 2 are as follows:

12) $[.110 \; x_2 + .053 \; x_{2A}]$ pounds of particulates

13) $[.040 \; x_2 + .016 \; x_{2A}]$ pounds of sulfur dioxide

14) $[.060 \; x_2 + .062 \; x_{2A}]$ pounds of hydrocarbons

15) $[.222 \; x_2 + .231 \; x_{2A}]$ pounds of nitrogen oxides

16) $[.180 \; x_2 + .187 \; x_{2A}]$ pounds of carbon monoxide

Observe that control method 2A is characterized by less emissions of particulates and sulfur dioxide than control method 2, but more emissions of other pollutants. Because of a lower BTU content of number one diesel fuel, approximately 4% more is required to replace number two fuel, thereby increasing emissions 4%. However, the lower sulfur content and cleaner nature of the substitute fuel result in a net reduction of sulfur dioxide and particulate emissions.

The emission factors in equations 12 through 16 correspond to elements of the E matrix in equation 4 above. The element E_{in} represents emissions of pollutant i associated with an activity unit of control method n. For any pollutant i, the summation of emission factor times activity level over all control methods is equal to total annual emissions of pollutant i in the airshed. The vector a in equation 4 is the vector of annual emission flows.

A relationship between annual emissions, expressed in pounds, a_i, and the ambient air quality, q_i, expressed in parts per million or micrograms per cubic meter is required. The work of Zimmer and Larsen (19) suggests, as a first approximation, a linear relationship of the form

17) $M_i \, a_i = q_i - b_i$

where q_i is the annual average air quality measurement for pollutant i at some central location in the airshed and b_i is the background concentration, a constant. The number M_i is assumed to be a meteorological constant, evaluated from observed data from prior years. The value of M_p for particulates in the St. Louis model was estimated from the following observations:

18) $q_p \, (1963) = \left\{ \begin{array}{l} \text{annual geometric mean for suspended} \\ \text{particulates at the St. Louis Civic Cen-} \\ \text{ter in 1963} \end{array} \right\} = 128 \ \mu g/m^3$
    [4, p. 36]

19) $a_p \, (1963) = \left\{ \begin{array}{l} \text{total particulate emissions in the} \\ \text{St. Louis airshed in 1963} \end{array} \right\} = 300{,}000{,}000 \text{ pounds}$
    [6, p. 454]

20) $b_p \qquad = \left\{ \begin{array}{l} \text{background concentrations for} \\ \text{Missouri} \end{array} \right\} = 31 \ \mu g/m^3$
    [20, p. 25]

Substituting these values in equation 17, the following approximate relationships are obtained for particulates.

21) $M_p = \dfrac{128 - 31}{300{,}000{,}000} = 325 \cdot 10^{-9}$

22) $325 \cdot 10^{-9} \cdot a_p = q_p - b_p$

This set of relationships is generalized for all pollutants in equation 5 of the matrix model. The matrix M is a (P x P) diagonal matrix whose non-zero elements are the meteorological constants M_i. The inequality constraint in equation 5 allows the obtained air quality measurements for any pollutant to be equal to or lower than (i.e. superior to) the specified level.[2]

---

[2] The St. Louis model included requirements on two additional pollutants, aldehydes and benzo(a) pyrene. The air quality goals for these pollutants were exceeded, i.e., they were "over controlled" as a by-product of meeting the air quality goals for the other five pollutants.

The assumption that the air quality measurements at a single receptor point represent air quality in the entire airshed is a simplification in the present model. The assumption that emissions from all sources, regardless of their location in the airshed, have an equal effect upon air quality at this receptor point is a further simplification. A more sophisticated model, proposed by Burton and Sanjour (2), incorporates a diffusion transport model and a sufficiently large set of receptor points that air quality standards can be presented in isopleths. The superiority of such a model is gained at the expense of great size and complexity. It should be noted that both models are more applicable to annual air quality averages than to short period maximums, which are an important dimension of air quality.

The results of running the above model for the St. Louis airshed with all N control methods may be reviewed by comparing Tables 1 and 2. Table 1 contains an emission inventory for the St. Louis airshed in 1975 in the absence of enforced abatement. Table 2 contains an emission inventory for the same set of sources, assuming that an efficient set of control method activity levels is instituted. This set of control method activity levels, denoted as x, is obtained by running equation 2 through 6 above with the value of q given in equation 1. The total minimized cost of control, z, is $35,350,000 per year. The last row of Table 2 contains the individual pollutant shadow prices. These are part of the MPS/360 program output and can be interpreted as the cost of abatement of the last pound of each pollutant.[3] They will be referred to at length, later in the paper.

Some observations on the solution of the St. Louis model should be made. Although the annual cost of abatement in the airshed in 1975 is $35,350,000 the required capital investment is $190,000,000.[4] With so large a commitment for fixed capital, there is the possibility of a conflict between short run and long run cost minimization. As emissions increase over time, it is likely that higher levels of abatement efficiency will become necessary. The least cost solution for 1975 may involve the purchase of equipment which will be inadequate by 1980. If it is too costly to obsolete or upgrade this equipment, the cost of air pollution control in 1980 will be higher than it might have been had previous capital expenditures not been made. This problem of long run efficiency can be investigated using a dynamic linear programming model.

---

[3] The shadow prices are the dual values for the null vector in equation 4.

[4] I am grateful to Professor M. Dohann of the California Institute of Technology for the suggestion that I determine the capital investment for air pollution control. This had not been included in my original program output. The $190,000,000 may be a conservative figure because it does not include capital outlays for additional natural gas pipeline. It should be noted that this capital investment produces, in addition to air pollution abatement, the recovery of an estimated $12,000,000 in by-products, the major portion of which is 650,000 tons of sulfuric acid. Although this acid is conservatively valued in the model, the fact that the estimated commercial production of sulfuric acid in the St. Louis area for 1975 is 1,500,000 tons (6, p. 392) may cause concern as to whether there will be a market for the additional by-product acid.

Table 1.

Emissions (in Millions of Pounds) in the St. Louis Airshed in 1975
in the Absence of Air Pollution Regulations

Category of Source	Particulate Emissions	Sulfur Dioxide Emissions	Hydro-carbon Emissions	Nitrogen Oxide Emissions	Carbon Monoxide Emissions	Cost of Control
Transportation	26	14	717	149	3,326	$0.
Combustion of Fuel Oil	4	55	1	33	n	0.
Combustion of Coal by Industrial and Commercial Users	48	192	2	31	8	0.
Combustion of Coal by Residential Users	22	54	5	4	23	0.
Combustion of Coal by Public Utilities	34	755	1	131	3	0.
Combustion of Natural and By-Product Gases	6	45	125	56	n	0.
Combustion of Refuse	57	2	293	2	92	0.
Industrial Processes	94	273	78	9	748	0.
Evaporation & Miscellaneous Minor Sources	9	0	298	0	0	0.
TOTAL	300	1,390	1,520	415	4,200	$0.

n = less than 500,000 pounds.

The fact that the solution for the St. Louis model is capital intensive may be related to a bias in the type of control methods considered. This is a bias in favor of what Teller calls "constant abatement" (14). There is a philosophy that it is wasteful to have constant abatement and that air quality standards can be met by "peak lopping." This is accomplished by forecasting periods of thermal inversion or other meteorological conditions adverse to atmospheric dispersion of pollutants and requiring abatment only at these times. Teller estimated that air quality standards for sulfur dioxide in the Nashville area could be achieved with forecasting abatement at one sixth the annual cost for constant abatement (14). Roach and Hewson report successful use of meteorological forecasting for air quality control in the Willamette Valley in Oregon. This is accomplished through the regulation of open burning (12). However, the Willamette Valley is an urban-rural region and, as can be seen in Table 1 above, the combustion of all refuse contributes only a small portion of the pollution in the St. Louis airshed.

Parsons and Croke, investigating the prospect for forecasting abatement in Chicago, found that the success of such a program would require the

Table 2.

Emissions (in Millions of Pounds) in the St. Louis Airshed in 1975 with
An Efficient Set of Air Pollution Control Methods

Category of Source	Particulate Emissions	Sulfur Dioxide Emissions	Hydro-carbon Emissions	Nitrogen Oxide Emissions	Carbon Monoxide Emissions	Cost of Control	Control Methods (A Summary)
Transportation	26	14	494	93	2,247	$16,540,000	Exhaust, crankcase & nitrogen oxide control systems for motor vehicles
Combustion of Fuel Oil	4	50	1	33	n	$ 100,000	1% sulfur content maximum on number 6 fuel oil
Combustion of Coal by Industrial and Commercial Users	4	120	4	26	6	$ 3,760,000	Conversion to gas, up-graded dust collection efficiency, low sulfur coal, and stack cleaning
Combustion of Coal by Residential Users	n	n	4	1	n	$ 1,390,000	Conversion to natural gas
Combustion of Coal by Public Utilities	2	98	1	90	2	$10,120,000	Stack cleaning processes upgraded dust collection efficiency, mine-mouth generation of electricity

Table 2 continued

Combustion of Natural and By-product Gases	6	45	125	56	n	0	No controls
Combustion of Refuse	28	1	123	2	42	$ 1,740,000	Discontinuation of open burning, conversion of burning dumps to land-fill
Industrial Processes	56	72	51	4	38	$ 1,640,000	Various process controls
Evaporation & Miscellaneous Minor Sources	9	0	192	0	0	$ 60,000	Replacement of cone roof gasoline storage tanks with floating roof tanks
TOTAL	135	400	995	305	2,335	$35,350,000	
Pollutant Shadow Prices	$.07748	$.02193	$.02476	$.32639	$.00428		

n = less than 500,000 pounds

availability of a substantial quantity of discretionary gas that could be temporarily substituted for coal. Since air pollution episodes in Chicago were found to occur on very cold days, Parsons and Croke doubted that discretionary gas would be available when it was needed. Another control method for implementing forecasting abatement is the temporary curtailment of industrial activity; this was adjudged to be economically infeasible for the Chicago area (11).

A risk associated with this approach to abatement is the failure to correctly predict appropriate weather conditions. McFarland, Barry and DeNardo investigated forecasting in the Pittsburgh area and reported that "at times the (air pollution) levels during non-advisory periods were higher than those during advisory periods." (9) Larsen has warned that, even with successful forecasting, "Varying emissions as a function of meteorology would result in borderline quality air much of the time." (8, p. 12) Norsworthy and Teller have an interesting response to Larsen's criticism; they are developing a linear programming model that combines both constant and forecasting abatement. Such a model may indicate how the superior air quality from constant abatement may be achieved with some of the cost advantages of forecasting abatement (10). Whether such a model is practical remains to be seen.

Determining Optimal Air Quality Standards

Let us consider a model in which air quality standards, q, and an efficient set of abatement activities, x are simultaneously determined. If we assume a given budget for abatement, z, we can use a pollution damage function, $H(q)$, which need not be expressed in dollars. The model has the form

23) Minimize $H(q)$

24) Subject to $\quad c\,x \quad\quad = z$

25) $\quad\quad\quad\quad\quad Ux \quad\quad = s$

26) $\quad\quad\quad\quad\quad Ex - a = 0$

27) $\quad\quad\quad q \quad - Ma = b$

28) $\quad\quad\quad q, \quad x \quad\quad \geqslant 0$

Consider this two-pollutant damage function:

29) $\quad H = .6\, q_s \cdot q_p$

where q_s is the maximum 24-hour ambient air concentration for sulfur dioxide in parts per million, during a pollution episode,

q_p is the maximum 24-hour ambient air concentration for suspended particulates in micrograms per cubic meter, during a pollution episode,

H is the number of excess deaths in a population of ten million, attributable to the two pollutants during a pollution episode.

Larsen, who discovered this relationship from data on pollution episodes in London and New York City, claims that the linear correlation coefficient between excess deaths and the product of the two pollutants is .98 (8, pp. 7-13).

This function was taken as a proxy for total damages attributable to annual average concentrations of sulfur dioxide and particulates in the St. Louis area. The budget constraint model, using the data from the St. Louis model for equations 24 through 27, letting z equal $35,350,000, which was the least cost solution from the preceeding model, and letting the air quality standards for carbon monoxide, hydrocarbons, and nitrogen oxides be the same levels, or better than, those specified in equation 1 above, determines optimal air quality standards for particulates and sulfur dioxide for the St. Louis airshed in 1975. The solution is illustrated graphically in Figure 1. The inside curve represents all possible air quality standards for sulfur dioxide and particulates obtainable in the St. Louis airshed in 1975, given total annual abatement expenditures of $35,350,000.[5]

Two iso-damage curves (there are an infinite number of them) are mapped in Figure 1. Note that the damage curve is convex, because of the synergistic effect of the two pollutants. Given a specific budget constraint, damage from sulfur dioxide and particulates is minimized at that point where the transformation curve touches an iso-damage curve which is furthest to the right. This occurs in the vicinity of the point, $\{q_s = .018$ ppm, $q_p = 80\mu g/m^3\}$, which represents the optimal air quality standards, given the standards assumed for the remaining pollutants. Associated with this solution is an efficient set (or sets) of control methods, $\overline{x}$.

Professor Donald Burlingame of Clarkson University and I have been working on a budget constraint model in which all five pollutant standards are variables. We are using the St. Louis abatement model and a linear damage function based on the work of Sterling, Pollack, and Weinham. The latter relates duration of hospital stay to pollutant levels and is based on Blue Cross records for the Los Angeles area (13).

A model in which q, x, and z are all determined has the form

30) Minimize $H(q) + cx$
31) Subject to $\quad Ux \quad = s$
32) $\qquad\qquad Ex - a = 0$
33) $\qquad\quad q \quad - Ma = b$
34) $\qquad\quad q, \quad x \quad \geqslant 0$

In this model, both $H(q)$ and cx are expressed in dollars. Kenneth Wieand of Washington University and I are experimenting with this kind of model. Wieand has spent several years analyzing the dollar cost of air pollution in

---

[5]This transformation curve is generated from equations 24 through 28, making "$q_s = M_s$ · $a_s + b_s$" the objective function, minimized for values of q_p, ranging from 69 $\mu g/m^3$ to 100 $\mu g/m^3$. In actuality, the curve consists of straight line segments. For example, the point, $\{q_s = .02$ ppm, $q_p = 75 \mu g/m^3\}$ lies on a straight line between the points $\{q_s = .0196$ ppm, $q_p = 75.64 \mu g/m^3\}$ and $\{q_s = .0201, q_p = 74.84 \mu g/m^3\}$.

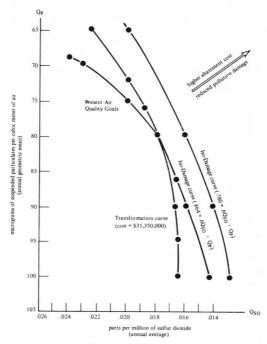

Figure 1: Iso-Damage and Iso-Cost Curves for Sulfur
Dioxide and Particulate Standards.

terms of reduced land values in the St. Louis area (16). Anderson and Crocker
have hypothesized that "many of the psychological and monetary costs
associated with living in a polluted environment such as additional cleaning
costs, loss of visibility, might be capitalized negatively into the value of sites
subject to pollution" (1, p. 2). Both Wieand and I, however, have reservations
about the validity of measuring air pollution damage through property values.

Michelson has developed a methodology for estimating the costs of
soiling attributable to suspended particulates. Using his findings, Wilson and
Minotte have implemented a benefit-cost model to determine q_p, x_p, and z_p
for the Washington, D. C. area (17). The subscript, p, here indicates that this
model deals with particulates only. Because air pollution control involves so
many parameters it may be necessary to start with one-pollutant models.
However, this can be a source of error, since control methods frequently affect
more than one pollutant. Figure 2, which is derived from the emissions and
control data for the St. Louis model (6), illustrates the shadow price for
achieving ambient air standards for particulates in the St. Louis airshed in
1975. Curve A is obtained by running the model for a range of q_p, without
requirements on the four remaining pollutants. For curve B, the requirements
on the remaining pollutants are those shown in equation 1 above. In this case,

113

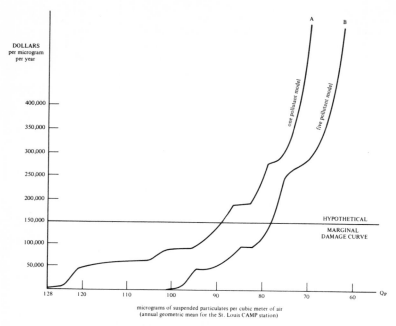

Figure 2: Shadow Price For Achieving Air Quality Goals
For Particulates in the St. Louis Airshed in 1975.

the multi-pollutant model has reduced the cost of particulate control by allocating part of control cost to the abatement of the remaining pollutants.

Assuming a constant marginal damage of $150,000 per $\mu g/m^3$ per year for suspended particulates, the optimal air quality standard should be $78\mu g/m^3$ according to the five-pollutant model, but only $89\mu g/m^3$ according to the one-pollutant model.[6] In this example, the one-pollutant model results in under abatement. To an extent, this problem also applies to the five-pollutant model which excludes not only a number of additional air pollutants but water and land pollution as well.

Figure 2 does indicate how air quality standards are a function not only of pollution damage but of abatement technology and the strategy for control, which in this case refers to the number of pollutant targets.

Reducing Source Magnitudes

The above models have assumed that the vector of source magnitudes, s, is fixed. A strategy for pollution control, which consists of charging sources

---

[6]Total damage plus abatement costs are minimized in the one pollutant model by reducing q_i to the point where the marginal damage equals the shadow price for control. If either function is discontinuous, abatement should cease just before the latter exceeds the former.

pollutant fees proportional to their emissions, could alter the vector, s, in favor of less polluting activities and at the same time encourage voluntary abatement. Such a strategy might be demonstrated for the St. Louis airshed, using data from the cost-effectiveness model above.

Assume that each source in the airshed in 1975 must pay pollutant fees equal to the shadow prices shown in Table 2. In the case of Source 2, the fees would total (see equations 12 through 16):

35) $(.110)(\$.07748) + (.040)(\$.02193) + (.060)(\$.02476) +$
$(.222)(\$.32639) + (.180)(\$.00428) = \$.084$ per gallon of
number two diesel fuel burned.

To minimize the sum of control costs plus pollutant fees, emitters would adopt control measures which cost less than the resulting reduction in fees. Using number one diesel fuel in buses (control method 2A) would reduce pollutant fees by

36) $(.057)(\$.07748) + (.024)(\$.02193) - (.002)(\$.02476) -$
$(.009)(\$.32639) - (.07)(\$.00428) = \$.002$ per gallon.

Since this is less than the control cost of \$ .015 per gallon, the bus companies would not use the substitute fuel.[7]

The significance of using the shadow prices in Table 2 as fees is that the resulting voluntary set of control methods will be the same as that summarized in Table 2. Not only will the same set of controls be adopted by emitter choice, but those sources which would have escaped regulation in the previous strategy, because control is relatively more costly for them, would pay pollutant fees instead. To the extent that these costs raise prices and reduce demand, such sources are controlled by reduced output.

The fee system is further demonstrated in Table 3 for selected sources. In the case of automobiles, a different fee for each make may be necessary. Although the problems of collecting differential fees at service stations seem insurmountable, there are advantages in leaving the choice of the type of control devices to motorists rather than the government. Assuming, for example, that the new device developed in California for burning natural gas in automobiles resulted in a total control and fee cost of four cents, only the market could determine if the penny savings per mile over the control system shown in Table 3 were enough to balance the loss in trunk space given over to the storage tanks.

For coal burning sources, a fee system might be more effective than codes on particulate emissions and regulations on sulfur content. Conversion

---

[7] David F. Merrion, Diesel Engine Division, General Motors Corporation, stated that pollution control for diesel engines will be accomplished through improved design. This control method was not included in the St. Louis model. (see *St. Louis Post-Dispatch*, "Design Called Key to Cleaner Diesel Engines," July 23, 1969, p. 5A)

Table 3

Selected Emitter Fees For the St. Louis Airshed in 1975[a]

Emission Source	Fees Without Abatement	Control Method	Cost of Control Method	Fees with Abatement
Diesel fuel burned in buses	$ .084 per gal.	Substitution of number one diesel fuel	$ .015 per gal.	$ .082 per gal.
Gasoline burned in motor vehicles	$ .07 per gal.	Factory installed crankcase, exhaust, and nitrogen oxides control devices	$ .03 per gal.	$ .02 per gal.
Industrial coal (3.1% sulfur) burned in spreader stokers, without air pollution control devices	$19.20 per ton	High efficiency cyclone collector	$ .65 per ton	$9.90 per ton
		Cyclone collector plus the substitution of 1.8% sulfur coal	$3.10 per ton	$8.20 per ton
		Conversion to natural gas	$6.50 per ton of coal replaced[b]	$1.60 per ton of coal replaced
Utility coal (3.1% sulfur) burned in the Sioux Power Plant. (A 50% reduction in effective emissions is credited to tall stacks)	$ 5.25 per ton	Stack cleaning process	$1.25 per ton	$3.25 per ton

116

Table 3 continued

		Baghouse filter	Catalytic combuster
Natural gas burned by residential users	$ .05 per thousand cubic ft. [c]		
Primary steel produced in electric arc furnaces	$ .80 per ton	$ .30	$ .01
Nitric acid produced	$18.00 per ton	$ .55	$ .90

a) Data on emission factors and control costs from R. E. Kohn, *A Linear Programming Model for Air Pollution Control in the St. Louis Airshed*, unpublished doctoral dissertation, Washington University, June 1969.

b) The $6.50 includes a scarcity premium of $3.55, which would presumably be paid as a tax to compensate future users of this scarce resource.

c) It is assumed that 1% of all natural gas sold leaks to the atmosphere through faulty pipes and regulator margins, as polluting hydrocarbons.

117

Table 4

Total Emission Fees Before and After Control in the St. Louis Airshed[a]

Category of Source	Fees Without Abatement	Control Methods (A Summary)	Cost of Control	Fees With Abatement
Transportation	$82,941,810	Exhaust, crankcase, and nitrogen oxide control systems for motor vehicles	$16,540,000	$54,524,370
Combustion of Fuel Oil	12,311,700	1% sulfur content maximum on number six fuel oil	100,000	12,202,050
Combustion of Coal by Industrial and Commercial Users	18,131,450	Conversion to gas, upgraded dust collection efficiency, low sulfur coal, stack cleaning	3,760,000	11,552,380
Combustion of Coal by Residential Users	4,416,580	Conversion to natural gas	1,390,000	425,430
Combustion of Coal by Public Utilities	61,986,160	Stack cleaning processes, upgraded dust collection efficiency, mine-mouth generation of electricity	10,120,000	31,712,520

Table 4 continued

Combustion of Natural and By-Product Gases	22,824,570	no controls	0	22,824,570
Combustion of Refuse	12,761,440	Discontinuation of open burning, conversion of burning dumps to landfill sites, etc.	1,740,000	6,069,390
Industrial Processes	21,340,240	Various process controls	1,640,000	8,648,800
Evaporation and Minor Sources	8,075,800	Replacement of cone roof tanks with floating roof tanks	60,000	5,451,240
TOTALS	$244,789,750	all control methods	$35,350,000	$153,410,750

a) This table is derived from Tables 1 and 2 above, using the shadow prices as pollutant fees.

from coal to natural gas, which might be unconstitutional if required by law, would be a likely option under a fee system, especially for the type of coal user illustrated in Table 3.

The largest users of coal in the St. Louis airshed are the electric utilities. If the fee system were adopted and the control and fee costs for the Sioux plant passed on to customers, the average price of electricity for residential users would rise from $ .0235 (25, p. 26) to $ .0253 per kilowatt hour.[8] Because of intense competition between the gas and electrical utilities to sell total energy systems, it is desirable that marginal pollution control costs be included in market calculations.

Table 3 should not give the impression that source fees can be calculated without difficulty.[9] Gerhardt has rightly warned of the difficulty in arriving at a fee structure (5, p. 10). However, by investing air pollution control with market payoffs, a fee system could provide an impetus to technological advances in abatement.

Vickrey, a proponent of the fee strategy, has noted that "effluent fees (could) provide a flow of revenues into the public treasury, (making) possible the abatement of other taxes and impositions that have a baneful rather than a beneficial effect on the efficiency with which the economy operates." (15, p. 2) Table 4 indicates that effluent fees in the St. Louis airshed in 1975, according to the demonstration model, would total $153,410,750. This may be compared to the total of $160,000,000 in general expenditures (other than for public education) in 1967, by the six county governments and the city of St. Louis, which together comprise the St. Louis airshed (23, 24). Although this tax revenue would be welcome, there is a risk that the burden of fees in addition to control costs might result in a migration of industry from the airshed. There is also the danger of undesirable distributional consequences. One-third of the fees will originate from transportation sources, a large portion from older vehicles for which control devices are especially costly.

Coase has written that "The Pigovian analysis (of social costs) shows us that it is possible to conceive of better worlds than the one in which we live. But the problem is to devise practical arrangements which will correct defects in one part of the system without causing more serious harm in other parts." (26) Whether a system of pollutant fees would be more satisfactory than a simple least-cost strategy I am unable to say. However, having observed for several years the frustrations to enforcement endured by regulatory agencies in the St. Louis airshed, I am inclined to view the fee strategy with increasing favor. Possibly some combination of fees and regulations, as suggested by Wolozin (18) could provide a viable strategy for air pollution control.

---

[8] One ton of coal produces 2500 KWH. (6, p. 299)

[9] If a given set of air quality standards were the objective of control, changes in the vector s should be anticipated. This would alter the optimal solution of the linear programming model as well as the shadow prices of the pollutants.

A Total Systems Approach

There is a body of opinion, held by Eckstein (3) and others, that air pollution is a sympton of inefficient economic and institutional structures. The cure, a reorganization of urban design, a restructuring of energy and transportation systems, etc., will produce superior air quality as part of the total utility pay-off. My model, in no way envisages the structural changes of the total system approach; instead it incorporates improvements into existing structures. Control strategy, however, can benefit by a broader environmental outlook. An air pollution model which incorporates constraints on water pollution, which reflects the increasing relative costs of solid waste disposal, which perhaps includes constraints on private vehicle miles driven to the central city, might bring us closer to the ideal of a total systems approach.

References

1. R. J. Anderson and T. D. Crocker, "The Site Value Approach to the Measurement of Economic Losses Due to Air Pollution," *APCA Paper 69-65*, New York City, June 22-26, 1969.

2. E. S. Burton, S. R. Peterkin, and W. Sanjour, "A Cost Effectiveness Approach to Urban Air Pollution Abatement," Presented at the 1968 Joint National Meeting of the Operations Research Society of America and the Institute of Management Sciences, San Francisco, May 3, 1968.

3. M. E. Eckstein, "Technology, Organizations, and the Social Aspects of the Control of Air Pollution," *APCA Paper 69-62*, New York City, June 22-26, 1969.

4. J. R. Farmer and J. D. Williams, *Interstate Air Pollution Study, Section III, Air Quality Measurements*, U. S. Public Health Service, 1966.

5. P. H. Gerhardt, "Costs and Incentives for Improving Air Quality," *APCA Paper 69-106*, New York City, June 22-26, 1969.

6. R. E. Kohn, *A Linear Programming Model for Air Pollution Control in the St. Louis Airshed*, Unpublished doctoral dissertation, Washington University, St. Louis, 1969.

7. T. C. Koopmans, *Three Essays on the State of Economic Science*, McGraw-Hill, Inc., New York, 1957.

8. R. I. Larsen, "Proceeding from Air Quality Criteria to Air Quality Standards and Emission Standards," *APCA Paper 69-210*, New York City, June 22-26, 1969.

9. D. G. McFarland, E. V. Barry, and J. W. DeNardo, "The Development of a Quantitative Objective Air Pollution Forecast System for Allegheny County Pennsylvania," *APCA Paper 69-76*, New York City, June 22-26, 1969.

10. J. R. Norsworthy and A. Teller, "The Evaluation of the Cost of Alternative Strategies for Air Pollution Control," *APCA Paper 69-172*, New York City, June 22-26, 1969.

11. D. O. Parsons and E. J. Croke, "An Economic Evaluation of Sulfur Dioxide Air Pollution Incident Control," *APCA Paper 69-20*, New York City, June 22-26, 1969.

12. M. D. Roach and E. W. Hewson, "Meteorology as a Tool in Air Pollution Control in the Willamette Valley, Oregon," *APCA Paper 69-110,* New York City, June 22-26, 1969.

13. T. D. Sterling, S. V. Pollack, and J. Weinham, "Measuring the Effect of Air Pollution on Urban Morbidity," *Arch. Environ. Health,* April 1969, *18,* 485-494.

14. A. Teller, "Air Pollution Abatement: Economic Rationality and Reality," *Daedalus,* Fall, 1967, 1082-1098.

15. W. Vickrey, "Theoretical and Practical Possibilities and Limitations of a Market Mechanism Approach to Air Pollution Control," presented at the Air Pollution Control Association Meetings, Cleveland, Ohio, June 11, 1967.

16. K. F. Wieand, "Effects of Air Pollution Upon Residential Location in St. Louis," *Mimeographed Doctoral Research Proposal,* Washington University, St. Louis, March, 1969.

17. R. D. Wilson and D. W. Minnotte, "A Cost Benefit Approach to Air Pollution Control," *J.A.P.C.A.,* May 1969, *19,* 303-308.

18. H. Wolozin, "The Economics of Air Pollution: Central Problems," *Law and Contemporary Problems,* Duke University School of Law, Durham, N.C., 227-238.

19. C. E. Zimmer and R. I. Larsen, "Calculating Air Quality and its Control," *J.A.P.C.A.,* December 1965, *15,* 565-572.

20. *Air Quality Data from the National Air Sampling Networks and Contributing State and Local Networks, 1964-65,* U.S. Public Health Service, 1966.

21. *Air Quality Data from the National Air Sampling Networks and Contributing State and Local Networks, 1966 Edition,* U.S. Public Health Service, Durham, N.C. 1968.

22. *Air Quality Standards and Air Pollution Control Regulations for the St. Louis Metropolitan Area,* Missouri Air Conservation Commission, Jefferson City, Missouri, 1967.

23. *City Government Finances in 1966-67, GF 67 – No. 2,* U.S. Department of Commerce, Bureau of the Census, Washington, D.C., 1968, p. 33.

24. *Census of Governments, 1967 Volume 4, Finances of County Governments,* U.S. Department of Commerce, Bureau of the Census, Washington, D.C., 1969, pp. 109, 110, 149, 150, 151.

25 *Annual Report 1968,* Union Electric Co., St. Louis.

26. R. H. Coase, "The Problem of Social Cost," *The Journal of Law and Economics,* October 1960, *3,* 1-44.

NOTE: *APCA Papers* were presented at the Sixty-Second Annual Meeting of the Air Pollution Control Association, June 22-26, 1969 in New York City. Copies may be purchased from the Air Pollution Control Association, 4400 Fifth Avenue, Pittsburgh, Pennsylvania, 15213.

Abatement Strategy and Air Quality Standards
Discussion

By John H. Niedercorn

Associate Professor of Economics and
Urban and Regional Planning at the
University of Southern California

I would like to congratulate Professor Kohn on his fine paper. It is an outstanding example of the creative application of economic theory to the solution of the nation's most threatening social problems. Professor Kohn outlines three possible approaches to air pollution abatement, the controls strategy, the fees strategy, and the total systems strategy.

The controls strategy involves the imposition of controls on sources of pollution in the manner that maximizes their effectiveness. Professor Kohn develops three mathematical models that could be used to attain this goal. The first achieves desired air quality levels at minimum total cost; the second minimizes damage subject to a cost constraint, and the third minimizes the sum of damage measured in dollars and the cost of abatement. Professor Kohn seems to assume that the results of a controls policy will be substantially the same irrespective of who bears the cost of control. However, it is clear that if the public foots the bill, the structure of production will probably not change significantly, but if the polluters pay the cost the use of pollutant emitting processes will be reduced.

Further research on the damage function is much to be desired. The form used in Equation (29) could be easily generalized by raising each of the independent variables to a positive power. A logarithmic transformation of the function could then be made so that its parameters could be estimated by regression analysis.

The fees strategy would provide a strong incentive for existing activities to reduce pollutant emissions, and would also stimulate the development of new production technologies designed to reduce them. However, the total fees suggested for the St. Louis airshed represent a very large sum of money. The author points out correctly that fees at such a level might cause industry to leave the area, and would constitute a hardship for automobile drivers, especially those in the lower income groups driving older cars. Nevertheless, if all the major cities imposed fees at the same time the above results would in general be beneficial.

At present industry is strongly attracted to metropolitan areas owing to various external economies associated with agglomeration and urbanization. However, the external diseconomies generated by air pollution are not reflected in the market. When internalized these diseconomies are quite likely in many cases to outweigh the economies. If firms in such circumstances relocate outside the airshed the overall level of economic welfare in the nation is increased. Similarly, under the fees strategy those who drive to work will have a strong incentive to make the socially desirable choice of relocating nearer their jobs, and thereby reducing traffic, highway costs, and pollution. Cities would tend to become more compact.

The controls and fees strategies are promising short and medium term devices for reducing pollution. However, it is clear that as the energy, water, and waste disposal needs of society continue to expand, a more radical long term approach will become necessary. Nothing less than a total systems strategy involving the design and construction of entire new, largely self-contained cities and the eventual abandonment of the old will insure the future of mankind on this earth. Two small prototypes of such cities are already on the drawing boards. Construction of both Seward's Success in Alaska and the Minnesota Experimental City is scheduled to begin soon.

VEHICULAR AIR POLLUTION:
VARIABLES INFLUENCING THE URBAN TRANSPORTATION SYSTEM

By Ibrahim M. Jammal
Assistant Professor
Department of Urban Planning
University of Washington

I. PERSPECTIVE

Recent trends in urban planning and urban development can be characterized by a renewed and intensive concern not only for the social and economic issues of the day, but also for the achievement of a better quality of the entire urban environment. This increased concern for environmental quality stems from the realization by planners, urbanologists and members of other related professions that urban areas are increasing their danger and risk of becoming places in which people could hardly live.

A comprehensive and rigorous definition of desirable environmental quality is still a very complex and difficult task. Discernible and tangible adverse effects on human, animal and plant life range from mental stresses and increased health hazards, to economic losses in real property and agricultural crops. Such effects are prompting intensive investigations of various components contributing to the deterioration of the environment.[1]

One of these major components is air pollution. Air, being a natural resource essential to human survival, is needed to sustain the kind of industrial world in which we presently live and for whatever activity in which we engage such as production, consumption, leisure, social interaction, and waste disposal. We seldom realize that, similar to other natural resources we use, the

---

*I wish to thank my colleague Dr. Anthony Tomazinis, University of Pennsylvania, for his valuable comments in my early thinking on this paper. His full contribution can be appreciated in a co-authored paper which developed from the present text. That paper entitled: "Community Variables Influencing the Urban Transportation System" will be published in the forthcoming volume *A Systems Approach to Environmental Quality: Vehicular Air Pollution and Public Policy.*

[1] e.g. *A Study of Pollution – Air: A Staff Report to the Committee on Public Works*, United states Senate, Sept. 1963.

supply of pollution-free air is not limitless.[2] Although we are able to compute the volume and weight of air necessary to carry on a variety of activities, we are still unable to gauge the absolute limits of the (pure) air supply likely to be available in the United States to man and his purposes over any one interval of time. The apparent ubiquity of air, its power to support life and property, its capacity to dispose of huge amounts of waste, its ability to renew itself through natural processes beyond the control of man, the tolerance of human beings and their ability to adapt to a wide but finite range of adverse environmental conditions as well as their pursuit of presumably more important ends, all these factors have tended to work against the development of a public attitude which regards air pollution as a serious problem, until adverse effects become tangible or some 'crisis' situation has developed and has been experienced.

The fixed sources of pollution are responsible for 39% of the total emissions, while the Internal Combustion Engine (ICE) is responsible for the largest share of contamination, namely 61%. Table 1 shows that in the United States, internal combustion vehicles produced in 1965 an average of about 75 million tons a year of pollutants, compared to about 22 million tons from manufacturing industry, 15 million tons from electric power generating plants, 7 million tons from space heating and 3 million tons from refuse disposal. Approximately a ton of air is required for every tankful of gasoline used by the average motor vehicle. A ton of air occupies a volume of 25,000 cubic feet. A billion gallons of fuel consumed annually by motor vehicles in the U.S. use 94 trillion cubic feet or 640 cubic miles of air. Some scientists would hold that the use of such figures, in the absence of some rigorously established magnitude of the total volume of "standard air," bears very close resemblance to "scare-tactics;" however, in light of the observed effects of contamination in large metropolitan areas, these figures point to the central role of the automobile in polluting the air and give some subjective magnitude of the problem we face.

Urban migration, increased standards of living, and increased multi-car ownerships reflect a preference of people for the car as their mode of transport. This means higher quantities of emitted pollutants concentrated in a very small proportion of the atmosphere, namely the atmosphere of major metropolitan urban areas which represent about 1.5% of the total area of the country. Table 1 indicates that in 1960, 60% of the total U.S. population was exposed to various degrees of air pollution and 24% of the population was exposed to major air pollution. We note from Table 2-a that the ICE vehicles

---

[2] "With most resources, whether renewable or nonrenewable, (man) is potentially able to modify to some degree their elemental and/or locational characteristics in order to suit his economic needs. But with air, man's actions, institutions, and artifacts must be modified. He is unable to adjust the winds to any appreciable extent; therefore, he must adjust himself to the vagaries of the air currents. His inability to face up to this fact appears to be root cause of the atmospheric-pollution problem" (p. 63). Thomas D. Crocker "The Structuring of Atmospheric-Pollution Control Systems" in *The Economics of Air Pollution*, Harold Wolozin ed., Morton & Co., N.Y., N.Y., 1966.

are the major source of pollution. Increasing urban concentrations and car ownership, Tables 2-b, c, d, e, can only point to a progressively worsening air contamination situation unless some measures of control are adopted and implemented immediately.

Studies investigating the control of air pollution have generally concentrated their efforts on stationary fixed sources. They investigated the nature and chemical composition of the pollutant, its characteristics of flow and dispersion and its effects on the quality of the ambient atmosphere. The location of the emitting sources relative to air currents and their impact on the variety of receptors was important in this research. Few studies were oriented to the economics of emissions and controls or their relationships to the form of the community where the emission occurs.[3] To studies of moving sources, especially motor vehicles, these considerations are relevant. We still need to investigate the context in which such vehicular air pollution occurs.

The use of the automobile is one of many modes of transportation in urban areas. Although pollution experts know about the performance of the IC engine and its contaminating emissions in the various phases of operation, no systematic control on an appropriate scale can be expected until we investigate the structure of urban areas and their movement systems in which contamination is generated.

Suggestions for controlling vehicular air pollution often revolve around a shift to mass transit, the compaction of cities, and the reduction of trip lengths. Such solutions are not as simple as they sound. The situation becomes much more complex as soon as one begins to operationalize these suggestions. Many considerations are involved such as the costs and benefits at the local and national levels ensuing from economic readjustment to new movement systems; questions of value preferences of people and styles of life, economic adaptability of investments in the existing environments which are automobile oriented, as well as political overtones reflecting anticipated economic and labor adjustment (not to say obsolescence). In investigating risks and costs which a community is willing to assume for its environmental health, Chambers has argued convincingly that any intervention in an existing "ecological" system can have repercussions well beyond the immediate focus of attention.[4]

---

[3]See for example:

Williams J. Pelle, *Annotated Bibliography on the Planning Aspects of Air Pollution Control.*Study for Northwestern, Illinois Planning Commission and U.S. Public Health Service, March 1965.

Harold Wolozin (ed.) *The Economics of Air Pollution.* W. W. Norton & Company, Inc., New York, N.Y., 1966.

Peter Rydell and Gretchen Schwartz, "Air Pollution & Urban Form" *Journal of AIP*, volume XXIV, No. 2, March 1968.

U.S. Department of H.E.W., *Air Pollution Publications, A Selected Bibliography with Abstracts 1966-1968*, Public Health Service Publication No. 979.

[4]Leslie A. Chambers, "Risks Versus Costs in Environmental Health," in Wolozin (ed.) *Economics of Air Pollution*, op. cit.

Table 1. Estimated[1] number of places with air pollution problems and population exposed to air pollution[2].

[1960 population in thousands]

Population class	All urban places		Major problem			Moderate problem			Minor problem			All problems		
	Number in class	Approximate population	Places Percent	Number	Approximate population	Places Percent	Number	Approximate population	Places Percent	Number	Approximate population	Places Percent	Number	Approximate population
Urban places:														
1,000,000 or more	5	17,500	100	5	17,500	0	0	0	0	0	0	100	5	17,500
500,000 to 1,000,000	16	11,100	70	11	7,800	30	5	3,300	0	0	0	100	16	11,100
250,000 to 500,000	30	10,700	45	13	4,800	45	14	4,800	10	3	1,100	100	30	10,700
100,000 to 250,000	81	11,600	25	20	2,900	50	40	5,800	25	21	2,900	100	81	11,600
50,000 to 100,000	201	13,800	20	40	2,800	35	70	4,800	45	91	6,200	100	201	13,800
25,000 to 50,000	432	14,900	10	43	1,500	25	108	3,700	45	194	6,700	80	345	11,900
10,000 to 25,000	1,134	17,600	8	91	1,400	20	227	3,500	37	420	6,500	65	738	11,400
5,000 to 10,000	1,394	9,800	3	42	290	12	168	1,200	35	487	3,400	50	697	4,890
2,500 to 5,000	2,152	7,600	2	43	150	10	215	760	28	602	2,100	40	860	3,010
Unincorporated parts of urbanized areas		9,900			3,800			2,300			2,300			8,400
SUBTOTAL	5,445	124,500	5	308	42,940	15	847	30,160	33	1,818	31,200	53	2,973	104,300
Urban & rural places under 2,500	14,345	11,100										[3]30	4,300	3,300
GRAND TOTAL	19,790	135,600	5	308	42,940	15	847	30,160	33	1,818	31,200	37	7,273	107,600
Percent of total U.S. population[4]		76			24			17			17			60

1. Accuracy of estimates not to be inferred from number of significant digits reported.
2. Urban places as defined by U. S. Department of Commerce, Bureau of the Census.
3. Problems are mostly minor.
4. Total U. S. population in 1960 was 179,323,000.

Source: A Study of Pollution — Air: A Staff Report to the Committee on Public Works, United States Senate, September 1963; as published in H. Wolozin (Ed.): *The Economics of Air Pollution*," W. W. Norton & Company Inc. N.Y., 1966.

Table 2a.

Source Distribution for Selected Air Pollutants in the United States – 1965**

Pollutant \ Source	Oxides of Nitrogen (10^6 tons/year)	%	Carbon Monoxide (10^6 tons/year)	%	Particulate (10^6 tons/year)	%	Oxides of Sulfur (10^6 tons/year)	%	Organics (10^6 tons/year)	%	Total By Source (10^6 tons/year)	%
Industrial Combustion & Processes	1.6	20	1.95	3	6.0	50	8.74	38	3.75	25	22.04	17.90
Refuse Disposal	0.65	< 1*	1.30	2	0.6	5	0.05	< 1*	1.05	7	3.05	2.50*
Transportation Int. Comb. Engines	3.2	40	59.8	92	1.8	15	0.46	2	1.75	65	75.01	61.00
Electric Power & Utilities	2.35	30*	0.36	< 1*	2.4	20	10.35	45	0.07	< 1*	15.49	12.60
Domestic & Commercial Space Heating	0.8	10	1.36	2.5	1.2	10	3.40	15*	0.38	2.5	7.41	6.00
Total Emissions by Pollutant, 10^6 Tons/Yr.	8	100*	65	100*	12	100	23	100*	15	100*	123	100

*Rounded Percentage.

**Totals for 1966 are: Industry – 16.8%; Refuse Disposal – 5.6%; Transportation (ICE) – 60.6%; Electric Power & Utilities – 14.1%; Space Heating – 5.6%. *Source:* The Sources of Air Pollution and Their Control, H.E.W., 1966.

Table 2b. Trend in Average Car Ownership for Various Sized Metropolitan Areas.

	Year	Present Ratio Cars/ 1000 Pop.	Year	Future Ratio Cars/ 1000 Pop.	% Increase
1,000,000 or more	1960	267	1980	330	23%
500,000 – 1,000,000	1960	310	1980	370	20%
250,000 – 500,000	1960	336	1980	390	16%
100,000 – 250,000	1960	329	1980	381	16%
Less than 100,000	1960	349	1980	380	8%

Source: Adapted from A. M. Voorhees
(Various metropolitan areas transportation studies).

Table 2c. 1960 and 1980 Population for Various Sized Metropolitan Areas.

	1960	1980	% Increase
1,000,000 or more	63,100,000	84,500,000	33%
500,000 – 1,000,000	17,200,000	22,400,000	30%
250,000 – 500,000	16,400,000	18,700,000	14%
100,000 – 250,000	14,300,000	16,300,000	14%
Less than 100,000	40,600,000	52,000,000	28%

Source: A. M. Voorhees (U. S. Bureau of the Census; Urban Land Institute).

Table 2d. Trend in Car Ownership for Various Sized Metropolitan Areas.

	1960	1980	% Increase
1,000,000 or more	16,800,000	27,800,000	65%
500,000 – 1,000,000	5,300,000	8,300,000	56%
250,000 – 500,000	5,500,000	7,300,000	33%
100,000 – 250,000	4,700,000	6,200,000	32%
Less than 100,000	14,100,000	19,800,000	40%

Source: A. M. Voorhees (Based on above tables).

Table 2e. Anticipated Urban Population Distribution

	Major Metropolitan Areas (Over 50,000)		All Urban Areas (Over 2500)
	Number	% of Population	% of Population
1960	78	48.3	67
1980	117	59.4	75
2000	145	65.9	85

Source: Jerome P. Pickard, *Metropolitization of the United States*
Urban Land Institute, Washington, D.C., 1959. U.S. Department of
Commerce, Bureau of the Census Data

One tends to initially ask the question: How permanent is the problem of vehicular air pollution? The answer really hinges on controlling contamination generated by the ICE, rather than on the concept of the automobile as a private vehicle. If fuel is changed or the characteristics of the engine changed to "clean" performance, then the problem of vehicular air pollution would be eliminated. A lot depends on the nature of improvement, since a substantial portion of the national economy is dependent upon the production of such vehicles, engines, spare parts and their maintenance; even phases of fuel production reach into the international realm.

We cannot disregard the fact that the automobile has exploded the city and released it from the traditional mile-distance constraint. It has made individual movement possible, although at a high cost for some. It has opened locational and preference choices heretofore unavailable and has become a way of life too entrenched in the culture of the people for us to expect it to easily change. Maybe the automobile (as we know it today) will change, but what the automobile introduced, namely, *the individual, flexible, comfortable (but not always the safest) form of movement*, will remain. The point of equilibrium between automobiles and mass transit[5] does not seem to be the economist's dollar point, but the perception of that point by the individual user through the values he has acquired by using the automobile, the way it has affected his experience, his range of choices and his preferences.

The preceding overview shows that inputs for solving problems of vehicular contamination are not only the function of the technology of the machine itself, but are intimately linked to the nature of the urban area, its population, its spatial organization and its dynamics of change. A premise of this paper is to accept the basic argument that "air is never completely pure." An air pollution problem tends to occur when man-made contaminants are discharged into the air rapidly or when they accumulate in such concentrations that the normal self cleansing or dispersive properties of the atmosphere cannot cope with them. As man cannot yet control the beneficial properties of the atmosphere, then air contamination can be considered as a differential in performance between sources of contamination and the controls for abating such contamination. Such a differential becomes a problem when it exceeds some implicit or explicit, subjective or objective levels of public acceptance and/or tolerance.

This paper does not concentrate on measuring the magnitude of that differential; it is primarily concerned with identifying the variables in an urban area which are responsible for, or influence the quantity of emitted vehicular contaminants and the possibility of controlling those variables.

The diagram in (Figure 1) is self explanatory and identifies the interrelated components whose nature and performance affect directly the magnitude of the vehicular contamination problem. Stationary sources and their control were included in the chart for the purpose of presenting a

[5] Elias, Gillies & Riemer, *Metropolis: Values in Conflict*, Wadsworth Publishing Col, Inc., Belmont, California, 1964. See Chapter 6, "The Automobile and Its Consequences."

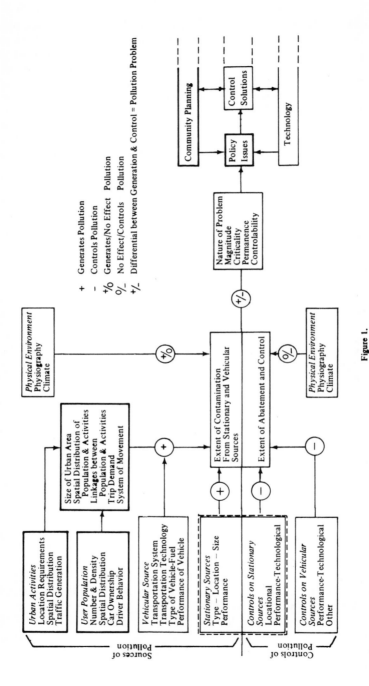

Figure 1.
Urban Area Components in Air Pollution Generation and Control

132

comprehensive picture; they have been the subject of intensive studies, the review of which is beyond the scope of this paper. It is also shown that the physiographic environment can be considered either as a source or a control to air pollution, depending upon whether its geography and climate would by design or the lack of it, help to disperse contaminants and mitigate their ill effects or contribute toward intensifying such effects.[6] The heavy edged boxes in Figure 1 indicate a range of urban area inputs which should be considered in vehicular contamination control. These inputs reflect the nature of urban activities and their spatial distribution as well as the distribution of population and the nature of linkages and interactions between people and activities.

II. SPATIAL ORGANIZATION OF URBAN AREAS

Understanding the spatial organization of the urban community is necessary to investigate the variables initial to the generation and control of vehicular air pollution. The type, patterns, and magnitude of vehicular movement is a reflection of the spatial interaction between people and activities in which they engage.

Although there exists no 'one' general theory of urban structure and urban growth, there have developed over time many contributions toward describing and explaining the organization and dynamics of urban areas.[7] These efforts draw on theories from land economics, sociology and ecology, but as such they are partial and reflect particular emphasis and points of view. What we will seek in the following discussion is to identify some general principles or factors which underly the spatial distribution in urban areas of people and activities and their relationship to the movement system which serves them.

Earlier contributions were based on imperical observation and generalized the spatial organization of people and activities in three concepts:[8] 1) the 'Concentric Zone' by Burgess, 2) the 'Sector' concept by Hoyt, and 3) the 'Multiple Nucleii' concept suggested by McKenzie, later expanded and refined

---

[6] See review article by Rydell & Schwartz, op. cit.

[7] For a general summary and review see Stuart Chapin: *Land Use Planning*, second edition, University of Illinois Press, Urbana 1965, Chapters 1, 2, 6.

[8] For a fuller theoretical discussion of these concepts, we refer to reader to:
Burgess, Ernest W. "The Growth of the City" in R. E. Park et al (eds.) *The City*, Chicago, University of Chicago Press, 1925.
Hoyt, Homer *The Structure and Growth of Residential Neighborhoods in American Cities*, Washington, D.C., FHA, 1939.
McKenzie, A. D. *The Metropolitan Community*, New York McGraw-Hill Co., 1933 and "The Scope of Human Ecology," Publications of American Society, XX (1926).
Harris, Chancy D. and Ullman, Edward "The Nature of Cities," The Annals of American Academy of Political and Social Sciences, Nov. 1945.
Ullman, Edward "The Nature of Cities Reconsidered," Papers and Proceedings of Regional Science Association, IX, Philadelphia, University of Pennsylvania, 1962.
Chapin, Stuart, op. cit., pp. 15-21.

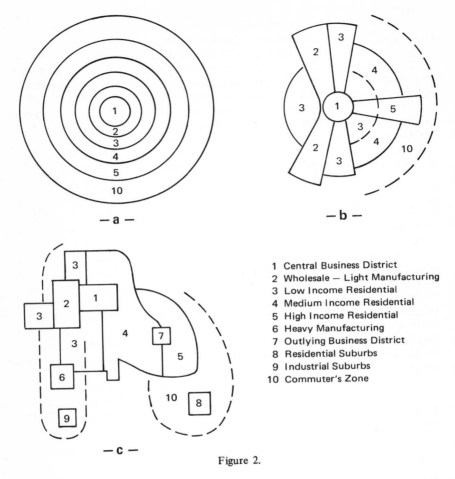

1 Central Business District
2 Wholesale — Light Manufacturing
3 Low Income Residential
4 Medium Income Residential
5 High Income Residential
6 Heavy Manufacturing
7 Outlying Business District
8 Residential Suburbs
9 Industrial Suburbs
10 Commuter's Zone

Figure 2.

by Harris and Ullman. (Figure 2) is a graphical representation of the three concepts.

Burgess simplifies the urban pattern in five concentric zones, each of which is quasi-homogenious in its character (Fig. 2-a). They are 'one center' oriented and grow outward, in ripples of invasion and succession, by the functions which seek locational advantages other than economic spatial centrality. Such functions should be able to afford paying for the resulting friction of space. Hoyt intended his theory to explain the distribution and growth of residential areas. He sees urban activities as organized in sectors radiating from a central core (Fig. 2-b). In each sector, the main transportation route is the spine around which urban activities locate; the growth of activities along these routes is considered to be 'more of the same' activity of each sector.

For McKenzie, "the distribution of human beings and human activities in urban areas are ecological processes dependent on the interplay of forces which affect a more or less conscious, or at any rate dynamic and vital relationship among the units comprising the aggregation."[9] Those ecological processes are unique in that they operate within a rather rigid structure base of establishment localities and movement channels of 'rather fixed spatial significance.' The ease of movement and the time-cost of distance have a strong effect on the ecological processes of 'specialization' and 'centralization' of activities and residences. In the growth of urban areas, these processes give rise to centers of various sizes and degrees of specialization. These centers would be classified according to: 1) size and importance of concentration, reflected in the magnitude and distribution of land values; 2) the dominant function of the center, such as work, business, or amusement; 3) the distance and the area of the zone of participation reflected in the location of the center at the point of easiest access to the participants. Although the spatial structural base is relatively fixed, such centers shift their location, over time, in response to trends in the distribution of the participating population. This shift is usually in order to improve access and minimize the friction of space between center and the largest number of actual and potential participants.

Harris and Ullman developed McKenzie's concepts further. They used the same classes of functions as Burgess and Hoyt and described the urban area in a more specific way as a 'multi nucleated' organization (Fig. 2-c). They hypothesized the rise of separate nuclei and differentiated districts as due to the combination of four factors. First, certain activities require specialized facilities and a certain level of accessibility obtainable only at their chosen location. Second, many activities are interdependent and tend to cluster together because they profit by cohesion. Third, some activities are detrimental to each other due to their particular requirements. Although these activities may have no affinity for each other, new centers develop to accommodate them. Fourth, different activities have varying abilities to afford high rents or land costs at their most desirable locations.

One can observe that, concerning the spatial relationships between activities, the three concepts have recognized the existence and dominance of a Central Business District where department stores, theatres, hotels, office buildings, banking houses, and other related businesses seek and can afford a central location. However, only in the multinucleated concept, do outlying business districts emerge. Contiguous or in close proximity to the CBD are wholesale, manufacturing and low income residential areas; better residences of the middle and higher income population are contiguous, but are further away from the CBD. Beyond the high income residential area is a commuter's zone.[10] In the multinuclei concept, we should note also that heavy

---

[9] op. cit., McKenzie "The Scope of Human Ecology."

[10] In Figures 2-b and c the commuter's zone is suggested by author and is not an original part of the concept.

manufacturing and industrial suburbs are still in close proximity to working men's homes and the wholesale high manufacturing districts. Their proximity suggests that the industrial suburbs are located on some existing transportation route serving the wholesale and industrial areas.

In these three location patterns of urban land uses, there is a tendency to minimize the cost of spatial separation between the residential locator and the activities in which he engages. This tendency becomes more apparent once we consider the nonresidential zones as places of employment and for provision of services. These tendencies are a function of the choices open to the locator. The choices include ability (given constraints of his income and his value-preference system) to make trade-offs between minimizing the cost of distance to his place of employment, his need for other services, and the proximity to other opportunities and amenities which are valuable to him. Aside from the fact that preferences also change with income, education and socio-cultural milieu, the residential choices open to the locator and his ability to exercise them (by making the appropriate trade-offs) increase as his income increases.

Within any urban area, the location of consumer services, wholesale, industrial, recreation, education, cultural and other related activities, is not as open to unlimited choices. Each has location and physical space requirements[11] which should be adjusted to constraints imposed by deter-minants of existing development as well as institutional controls governing the use of the land. Where the four factors advanced by Harris and Ullman would explain WHY the centers of activity emerge, the location and space requirements elaborated by Chapin would explain some of the physical factors governing WHERE they emerge.

Overcoming the distance between people and facilities and minimizing the relative cost of that distance is one of the major objectives of the locator. Haig and later Ratcliff argued that the efficiency in the pattern of the city is inversely proportional to the aggregate "costs of friction of space," i.e. the cost of overcoming distance. Taken to its logical conclusion, maximum efficiency is theoretically one where transportation cost is equal to zero,[12] meaning that all people and all facilities are located at one point.

The physical and cultural impossibility of such concentration coupled with the relative importance of other various conveniences valuable to the people, as well as the competitive process of location make spatial separation

---

[11] Chapin, op. cit., Chapters 10 and 11.
For a more rigorous discussion of an economist's point of view, see: Edgar M. Hoover, "The Evolving Form and Organization of the Metropolis" in *Issues in Urban Economics*, Harvey S. Perloff and Lowdon Wingo, Jr., (eds.), RFF-John Hopkins Press, 1968, pp. 237-284.

[12] Richard Ratcliff, "The Dynamics of Efficiency in the Locational Distribution of Urban Activities." *Readings in Urban Geography*, Mayer and Kohn (eds.), University of Chicago Press, 1959, pp. 299-324.

of people and facilities inevitable. Haig and Ratcliff's hypothesis of urban structure and growth is that:

> The city is a nucleated organism oriented to the center, and that the natural force which has produced this conformation as an adaptation to man's needs and preferences is the disutility of space giving rise to a constant effort to minimize the cost of friction. Concentration rather than diffusion is the natural product. Decentralization of certain lower-intensity land uses which are forced out of the center by higher-intensity replacements is a normal and continuing process of urban dynamics which enhances efficiency.[13]

Although still hypothetical, the conceptual framework developed by Gutenberg[14] is relevant to our concerns and merits some discussion. In his framework for urban structure he emphasizes the interplay between the location of urban activities and transport efficiency. The organizing principle of structure is to overcome the sum of all distances between persons and facilities. As not all persons are mobile, nor all facilities distributable, total distance can be overcome by the distribution of facilities in close proximity to users and/or making facilities accessible to people, i.e., transporting people to facilities. The function of the transportation system is to overcome residual distance between people and undistributed facilities after the distributable ones have been localized. Minimizing total distance implies minimizing residual distance; this is a function of the geometry of distribution and implies the need for spatial consolidation and centralization of undistributed facilities:

> This fact gives functional meaning to the core of the city or metropolis and to the radial shape of transportation serving it.[15]

The different degrees in the mobility of persons as well as differences in degree of locational, functional concentration of activities, accessibility requirements and service radii begin to indicate a hierarchy in the structure and explain the emergence of centers of varying sizes and the corresponding hierarchy in the transportation system serving them.

Density at any place is a function of the number of locators desiring that location, the efficiency of its access to distributed and undistributed facilities, and its ability to substitute for the central location, i.e., the ability of the locator to bear the cost of overcoming residual distance. The density gradient slopes negatively away from central points of highest access as residual

---

[13] *Ibid.*

[14] Albert Z. Guttenberg, "Urban Structure and Urban Growth" *AIPJ* volume XXVI, Number 2, May 1960, pp. 104-110. The text and (Figure 3) are an adaptation from Guttenberg's article cited above.

[15] *Ibid.*

distance increases and the substitutability of outlying points to the more central ones declines. As transportation efficiency decreases, so do economic and physical densities; however, the relationship between the two is not necessarily a direct one. When both densities correspond, they are usually the function of market governed forces; they reflect the scarcity of location opportunities, the linkages necessary for the operation of the locators and some cost/benefit evaluation of their operation at that location. When economic and physical densities do not correspond, physical density usually reflects the uneven distribution of those who seek in their location more of the intangible, non-empirical and hard to measure considerations such as locational symbolism, social identity, and certain amenities of the environment.[16]

Gutenberg considers growth as 'an increase in size and an adjustment to that size.' The adjustment occurs on elements of structure: viz. distribution and nodality of facilities, their hierarchy, the makeup of the transportation system and its efficiency. Underlying growth is a basic relationship between the spatial extent of a community and its structure. If major classes of activity such as work, play and residence are separated on a regional scale, the extent of the urban area is limited by the ability of the people to reach and move between the locations of such activities. On the other hand, if the same classes of activities are distributed at a local level, there would be no imposed limit on the size of the urban area. In the first case, daily routine might require a person to traverse the urban area several times over; whereas in the second case, the need for movement is minimized. In the first case, growth over time requires structural change in the form either of faster or more far-reaching transportation, or of new regional centers. Hence most changes and adjustments are dependent on the efficiency of transportation; for efficiency governs the substitutability of outlying locations for the more central areas. If efficiency is equal in all directions, the prevailing structure will be circular around a major center of undistributed facilities. Superior efficiency along transport-routes radiating from a major center and providing continuous access would result in sector-like development with ridges of high density development along these routes. If the access is discontinuous, the result would be a weblike structure in its centers of distributed facilities occurring at points of high access; e.g., the intersection of radial and circumferential highways. According to the continuity and degree of access, ridges of higher density could develop along radial and circumferential links.

The interaction between people and facilities could, in the broad sense, be considered as linkages. Some linkages such as flows of money, communication of information are not physical. Other linkages require the physical movement of people and goods to carry on their activities. The role of linkages in the relation of urban activities to the generation of traffic movement and the effects that transportation systems have on the location of land uses has

---

[16] See Walter Firey, *Land Use In Central Boston*, Cambridge, Harvard University Press, 1947.

Figure 3.

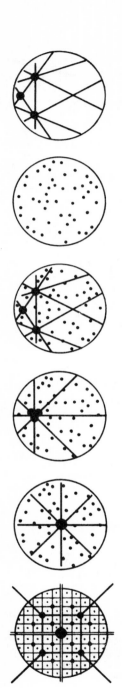

Transportation overcomes distance between people and facilities.

Distribution overcomes distance between people and facilities.

Not all people are mobile, and not all facilities are distributable. Combination between distributed facilities, undistributed facilities and transportation, overcomes total distance.

Consolidation of undistributed facilities reduces total distance.

Consolidation and centralization of undistributed facilities minimize total distance.

There develops a hierarchy of centers by size and function, which corresponds to a hierarchy of transportation channels serving those centers. Distributed facilities become the basis for local organization of human activities; undistributed facilities + transportation system become the basis for regional organization of human activities.

139

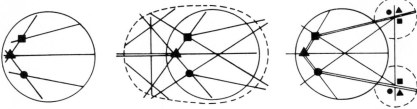

Growth is an increase in size & an adjustment to that size: regional separation of activities limits size in as far as people could reach those facilities; new transportation facilities or new regional centers may be necessary. But if major activities are centralized and distributed at a local level, there is little imposed limit on the size of the urban area.

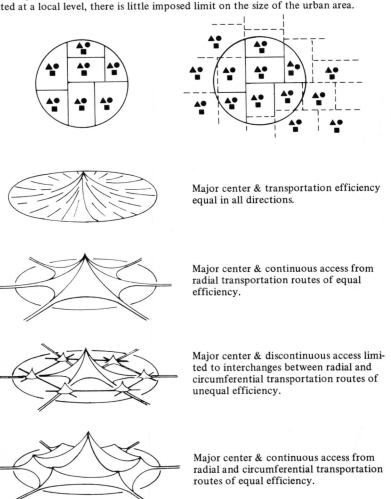

Major center & transportation efficiency equal in all directions.

Major center & continuous access from radial transportation routes of equal efficiency.

Major center & discontinuous access limited to interchanges between radial and circumferential transportation routes of unequal efficiency.

Major center & continuous access from radial and circumferential transportation routes of equal efficiency.

been investigated by Mitchel and Rapkin[17] in 1954. They suggested a framework for a systematic analysis of these interdependencies. Of special interest to this paper is their discussion of the influence of movement on land use patterns[18] in which they are concerned with how movement systems relate to the location of establishments and movement channels, what are the forces that make for changing movement requisites and how they influence the pattern of uses. Their discussion is basic. Any restatement or paraphrasing would not bring out all of its pertinent points, therefore, we encourage the reader to read the full discussion. Here is the textual reproduction of the summary and conclusion:

> The processes by which various land uses sort themselves out in the urban pattern are in a large measure a function of the movement requirements of establishments. Specialization of urban activities makes it necessary for establishments and their members to communicate with each other, and consequently there is a pervasive tendency for establishments to make accessibility a major locational consideration. The pattern of land uses is thus a large dependent system in which choice of location of an establishment is made in terms of spatial distribution of others with which it interacts. For some this means access to the largest number of persons, firms, or households — a central location; for others it means convenience in regard to an inexpensive channel of goods-movements; and in still other cases it means actual proximity. The tendency for certain kinds of establishments to seek proximity is described by the concept of linkage which aids in understanding the clustering of like and unlike establishments that characterize land use patterns. The concept of linkage also enables the translation of systems of social and economic action into ecological terms.
>
> Shifts in locational pattern may come about as a result of changes in the movement requirements that are due to alterations in the internal activities of an establishment, to response of an establishment to shifts in the locations of others with which it is linked, or to differential rates of growth among establishments of various types. Changes in the movement requirements of establishment or area will operate as a force to alter existing channels of movement. This relationship between the locational distribution of establishments and movement channels is reciprocal in character. A changing land use pattern will generate the need for additional physical channels of movement and new or changed

---

[17]Robert Mitchel and Chester Rapkin, *Urban Traffic: A Function of Land Use*, Columbia University New York, N.Y., 1954.

[18]*Ibid.*, Chapter VII, pp. 104-133.

traffic facilities will in turn encourage change in the existing distribution of establishments. This interpenetration of influences is significant in the attempts to rationalize the movement structure of an urban area. With further understanding it may also provide a basis for employment traffic channels as an instrument to secure a more efficient distribution of urban land use.[19]

In the preceding discussion we have dealt with basic concepts of urban structure which relate spatial patterns of land use to its transportation system. Recent work on models for transportation systems and for the allocation of land use should not be overlooked.[20] These efforts stem from those basic concepts but they are much more sophisticated in their development of rigorous analytical frameworks and their formulation of relationships. We could generalize by saying that they are concerned with simulating the locational decision of establishments and/or the performance and characteristics of transportation systems. Although the aggregate of locational decisions and performance of activity systems could provide generalizations about the structure of the urban area, the efforts of these models are not directed toward this end yet. The interested reader is encouraged to become familar with these models and attempt to draw on their conclusions in any further investigation of the role of urban structure in controlling vehicular contamination.

III. INPUTS IN CONTROL OF VEHICULAR AIR POLLUTION

The mechanical performance of the current internal combustion engine is known and the amount of pollutants emitted in the phases of acceleration, cruising, and deceleration are accurately measured. The quanity of pollutant emitted is a function of 1) rate of emissions in each phase; 2) the frequency of each phase per unit length of trip; 3) the length of the trip, and 4) the frequency of the trip. If we make the strong assumption that the source of vehicular pollution is not open to technological control, i.e., that we cannot

---

[19]*Ibid.*, pp. 132-133.

[20]Reference is especially made to the following:
1. Ira S. Lowry, *Model of Metropolis*, Memorandum RM-4035-RC, Rand Corporation, Santa Monica, California, August, 1965.
2. ——————. Seven Models of Urban Development: A Structural Comparison, Rand Paper P-3673, Rand Corporation, Santa Monica, September, 1967.
3. Melvin Webber (ed.) *Explorations in Urban Structure*, Philadelphia, University of Pennsylvania Press, 1964. See articles by Donald F. Foley, "An Approach to Metropolitan Spatial Structure" and Weber's "The Urban Place and Nonplace Urban Realm."
4. Stuart Chapin and Henry Hightwoer, *Household Activity Systems – A Pilot Investigation*, Center for Urban and Regional studies, Institute for Research in Social Science, University of North Carolina, Chapel Hill, May 1966.
5. George Hemmens, *The Structure of Urban Activity Linkages*, Center for Urban and Regional studies, Institute for Research in Social Science, University of North Carolina, Chapel Hill, September, 1966.

alter the performance of the engine, then the only way open to us would be to alter the four variables affecting the quantity of emissions. Thus, in terms of urban structure, several relationships which may affect the quantity of emitted pollutants could be discussed:

1. Relationship of the trip length to the shape of the city.
2. Relationship of the trip length to the size of the city.
3. Altering the relationship between transportation network and the location of urban activities.
4. Improving the performance of the transportation network.
5. Improving the relationship between the location of urban development and transportation network and the meterology of the area.

1. *Relation of trip length to the shape of the city.*

The studies presented here[21] are two examples of theoretical work which investigate the geometrical relationship between the shape of the city and the length of trip.

Circular and rectangular shaped cities were assumed. These cities are inhomogeneous in that residences and workplaces are not evenly distributed over the city area. Their simplified structure consists of an inner business district where all work places are located and an outer concentric residential zone where all residences are located.

In the circular shape, the Business district has a radius = a and the outer limits of the city have a radius = b. A person can travel from his home to his work place in any one of five ways (Fig. 4-a).

1. *Direct*: by going in a straight line from his home to his destination.
2. *Ring*: by going along a radial until he reaches the inner circle, then along the circumference until he reaches the radial on which his destination lies and finally straight to his destination.
3. *Rectangular*: by going along one of the many shortest routes from his home to his work place, utilizing routes that are either parallel or perpendicular to a particular diameter of the circle.
4. *Radial*: by going along a radial to the center of the city and then along another radial to his destination.
5. *Arc-radial*: by going along a circle concentric with his home until he reaches a point on the same radial as his destination and then straight along the radial to his destination.

Tan derives two graphics from his calculations (Figure 4-b, c) showing the relationship between the overall radius of the city (as a function of the radius of CBD, average distance travelled and average travel time). His

---

[21]T. Tan, "Rectangular Routing Systems in a Model City," Third conference of the Australian Road Research Board, Sydney, 1966.
T. Tan, "Road Networks in an Expanding Circular City," *Operations Research*, Vol. 1A, Number 4, July-August, 1966.
Frank A. Haight, "Source Probability Distributions Associated with Commuter Level in a Homogeneous Circular City," *TORSA*, Vol. 12, 1964.

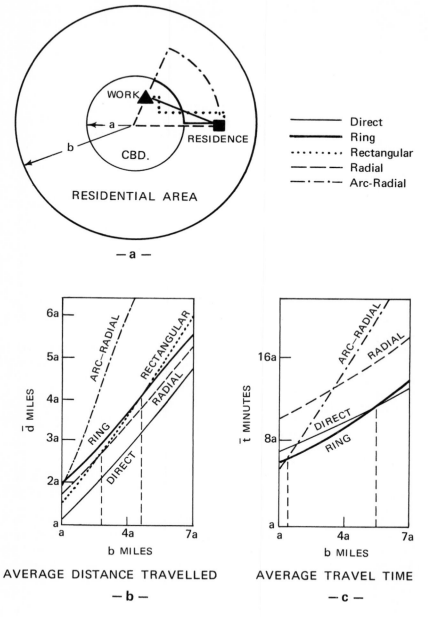

WORK

RESIDENCE

a

b

CBD.

RESIDENTIAL AREA

— a —

Direct
Ring
Rectangular
Radial
Arc-Radial

ARC—RADIAL

RECTANGULAR

RADIAL

RING

DIRECT

d̄ MILES

6a

5a

4a

3a

2a

a

a   4a   7a

b MILES

AVERAGE DISTANCE TRAVELLED

— b —

ARC—RADIAL

RADIAL

DIRECT

RING

t̄ MINUTES

16a

8a

a

a   4a   7a

b MILES

AVERAGE TRAVEL TIME

— c —

Figure 4.

144

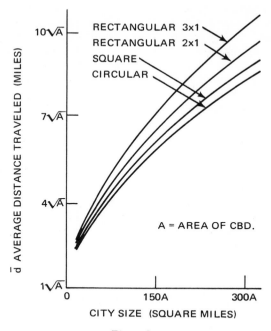

Figure 5.

conclusions are: 1) that as the city grows, the distance in miles increases, but there are variations; 2) the common rectangular system is useful for cities with compact residential zones and that beyond a certain critical point, the rectangular system should be changed to a radial or a ring system; and 3) a minimum average travel time is obtained with the Arc-radial system when b < 1.4a miles with the Ring system for the range 1.4a miles < b < 5.8a miles and with the Direct road system for b > 5.8a miles.

In the second of his studies, Tan proceeds to compare the average distance traveled from a point in the residential zone to a point in the CBD using the current rectangular system with facilities of the same area, but of various shapes.(Figure 5) is a resulting graph showing that the circular shape has the shortest traveled distance and that the differential between the shapes increases as the area of city increases in relationship to the size of the CBD. ·

Focusing his analysis on the square inhomogeneous city, he finds that rectangular routing system yields the shortest average trip length from a point in the residential area to a point in the CBD up to the point where the total area of the city is 64.2 the size of the CBD. Beyond this point the Radial Rectangular system is better. When the total size reaches 292.7 the size of the CBD, the mass transit system yields the shortest distance. Similarly, in terms of average time traveled, (assuming the following mph: residential 30, freeways or fast mass transit 40, ring roads 30, CBD streets 10), when the

145

Rectangular system is compared with the Radial-rectangular and mass transit systems, the total city size has critical values at 100A and 350A respectively, where A equals the size of the CBD. In both measures, however, the mass transit system seemed to be longer in length of trip measured either by miles or minutes. It was also shown that when ring roads were added to the radial rectangular system, the results were to reduce average travel time by about 30% from that of the mass transit system and 15% from the radial rectangular system for all sizes of square cities.

2. *Relationship of trip length to the size of the city*

The size of the city is one of the few variables which affects the length of the trip. Size could be considered either as the population number or as the city area. The relationship between these two aspects is one of density. If we keep the area of the city constant, density increases with population; and if we keep the population constant, the area decreases with increased density. In the second case the city becomes more compact. The morphology of the American city shows that almost invariably the area of the city increases with its increase in population.

In discussing the variation in the shape of the city, a theoretical geometrical relationship was introduced between the size (area) of the city and the length of trip. However, the assumptions of a homogeneous CBD surrounded by a homogeneous residential ring are oversimplifications applicable to a limited class of small towns. In the real world, this homogenity does not hold, for there arise various degrees of diffusion and intermixing between residences, services and employment centers. Empirical studies[22] investigating variables responsible for trip length variations measure this diffusion by the availability of jobs within any one ring around the CBD. This concentration, identified as "City Centrality," was found to decrease as the size and area of the city increased. Small cities exhibit more dependency on their central areas and are found to experience a marked and systematic increase in the length of their trips as the distance between the area of trip origin and the CBD increased. As the city gets larger, its CBD becomes less of a unique regional center; many other subcenters and satellites develop in outlying areas, thus improving the accessibility of residents to services and jobs. These impacts of subcenters on reducing traffic confluence to the CBD on tending to minimize home-work separation and, hence, on reducing the length and cost of the auto trip, have been first identified by Foley and Carroll[23] in the early fifties; the empirical study of the trip length variations

[22] e.g. Tomazinis and Gabbour, *Trip Length Variations in Urban Areas,* Institute for Environmental Studies, University of Pennsylvania, Philadelphia, June, 1966.
Also, Tomazinis, Hill and Jammal, *Trip Length Analysis Within the Philadelphia Metropolitan Region*, Institute for Urban Studies, University of Pennsylvania, March, 1964.

[23] J. Donald Foley: "The Daily Movement of Population into Central Business Districts," *American Socialogical Review.*
J. Douglas Carroll, Jr., "The Relation of Homes to Work Places and the Spatial Patterns of Cities," *Social Forces.*

in five cities of different sizes substantiate them. Although this study (Tomazinis and Gabbour) cautions against generalizations due to difficulties in the choice and measurement of variables, I feel that its findings are indicative of tendencies relevant to our discussion:

1. In small cities the length of trip consistently increases as the distance of the trip origin from the center of the city increases.
2. In large size cities the length of auto trips does not seem to be affected by the distance to the CBD. Autotrips, at times, even become shorter as distance increases due to the emergence of outlying regional centers.
3. Transit trips appear to be more affected by distance variations than autotrips in all size cities.
4. Households of higher income undertake longer trips than households of lower income. This is more true for total home origin trips than for auto work trips.
5. Income appears to be a relatively more influential factor affecting trip length in small size cities than in large urban areas.
6. Smaller cities demonstrate a greater concentration of economic activities within and on the periphery of the CBD. The association between trip length variations and the size of urban areas appears to be a consequence of this concentration of nonresidential activities in the CBD rather than a result of the size of the cities per se.
7. In small size cities the order of statistical influence is distance of origin from CBD, trip density (expressing the density of development), and income. In larger cities the incorporation of income does not seem to improve the statistical relationships.
8. Trip length, measured in miles, for home to work trips appears to decrease with the increase of the availability of expressway facilities, especially the ones from high and middle income households.

In a paper on "Variables Affecting Urban Traffic Behavior," A. M. Voorhees presents some data (Table 3) on major factors influencing trip length. The scatter diagrams (Figures 6-A, B, C) are for work trips; they represent, respectively, the relationship between the population size of the urban area and average trip length, average trip duration and average network speed. It should be noted that unlike the Tomazinis-Gabbour study, the Voorhees data does not report on density distribution of population, which relates the size of the city to its compactness or spatial spread, nor does it relate the length of trip to distance of trip origin from the center of the urban area.

By visual inspection, diagrams 6A and 6B suggest a positive convex function. If we assume this function to be correct and to fit the data, we may say that the length of the work trip, measured in miles or in minutes, increases with the size of the urban area but at a decreasing rate, tending to flatten in urban areas over 1,000,000 population. Diagram 6C suggests a linear function, parallel to the axis, and indicates that the average speed of the network tends

Table 3.

Work Trip Length Characteristics and Major Influencing Factors*

City	Population (Thousands)	Average Trip Duration (Min.)	Average Trip Length (Miles)	Average Network Speed (mph.)
1. Los Angeles	6,489	16.8	8.7	31.0
2. Philadelphia	3,635	20.1	7.2	21.5
3. Washington, D.C.	1,808	14.1	5.9	24.7
4. Pittsburgh	1,804	12.6	4.2	20.7
5. Baltimore	1,419	16.7	7.0	24.6
6. Minneapolis-St. Paul	1,377	12.5	5.1	24.5
7. New Orleans	845	9.1	3.0	20.2
8. Providence	685	14.6	–	–
9. Fort Worth	503	7.1	8.1	30.9
10 Lackawanna Luzerne[a]	453	19.2	8.7	27.1
11. Broward County	440	13.7	–	–
12. Ottawa-Hull	406	12.6	3.1	25.2
13. Nashville	347	10.8	5.4	30.0
14. Edmonton	336	11.6	5.8	30.0
15. Worcester	281	10.9[b]	4.9	27.3
16. Virginia Peninsula[a]	277	11.5[b]	6.3	33.0
17. Knoxville	258	9.4[c]	–	–
18. Davenport	227	7.7	3.2	24.9
19. Charlotte	210	11.0	5.5	30.0
20. Chattanooga	205	10.8	5.4	30.0
21. Eric	177	9.4	3.4	21.7
22. Waterbury	142	10.1	5.9	35.0
23. Springfield, Ill.	134	7.5	3.6	29.2
24. Pensacola	128	8.7	4.4	30.3
25. Regina	127	8.0	3.3	24.5
26. Greensboro	123	8.9	4.3	29.0
27. Lexington	112	9.1	5.7	35.0
28. Springfield, Mo.	110	8.4	–	–
29. Altoona	103	11.1	3.1	27.2
30. St. Catharines	99	13.6	–	–
31. Sioux Falls	67	7.0	2.9	24.8
32. Tallahassee	48	7.3	3.7	30.4
33. Hutchison	38	6.1	2.0	19.2
34. Beloit	33	6.7	2.9	25.9

*Source: Alan Voorees "Variables Affecting Urban Traffic Behavior," January 1968. Qualifications of data apply:

[a] External trips included in computing the average.

[b] Total persons trips.

[c] Average of blue and white collar average work trip lengths.

148

Figure 6,A

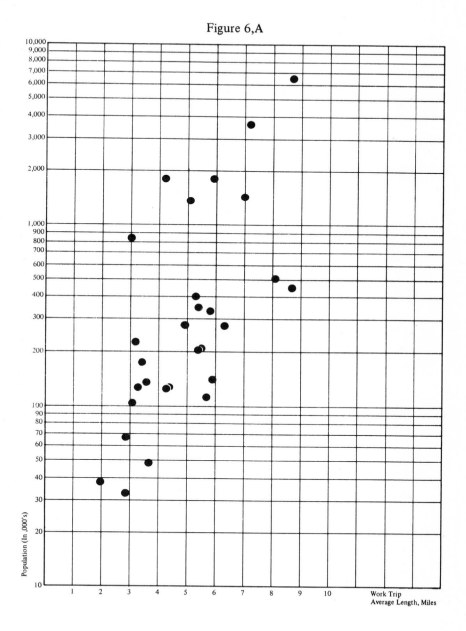

Figure 6,B

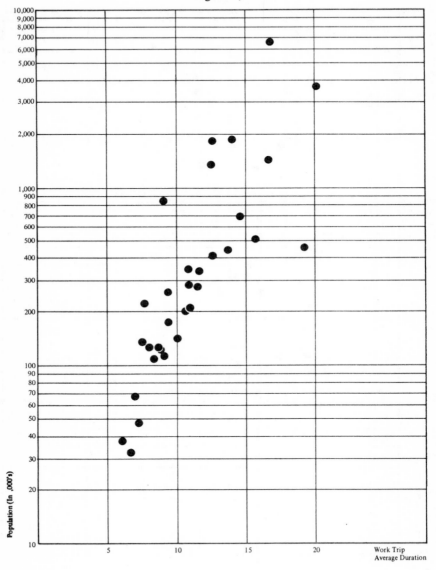

Figure 6,C

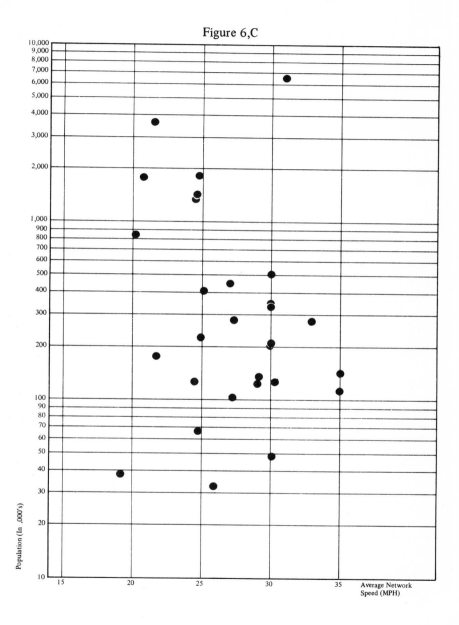

not to vary with increase in the size of the urban area. The import of this relationship will become apparent when we consider improving the performance of the transportation network later in the paper. The above suggestions are speculative; their validity depends on further empirical data and more rigorous statistical analysis.

3. *Altering the relation between transportation networks and the location of urban activities.*

In the earlier discussion of the spatial structure of urban areas, we have seen the close interdependence between the location of urban activities and the transportation system serving them. In this interpendence, the transportation system is expected to provide needed accessibility and accommodate the demand for trips (present as well as expected), given a known technology, some desirable level of performance, some feasible range of cost and some reasonable expectation of future change. On the other hand, urban activities seek, among other objectives, that location which is central or proximate to linked activities and where the level of accessibility is most suitable to conduct a profitable operation. Whether they are able to secure such a location is dependent on their ability to pay for it; their inability to pay is demonstrated by their choice of 'second and third best' locations where the decreasing accessibility is reflected in a lowered centrality and an increased transportation cost.

It is in the dynamics of growth that problems of adjustment become apparent. There exists a differential in the rate and ability of either system to adjust to change introduced in the other system. Adaptation is a costly process, justified when its present and future benefits outweigh its immediate and opportunity costs given some acceptable measure of value. The huge amounts of funds sunk in the physical plant is a major problem. The ability of the system to adapt to change seems to depend on whether or not such costs can be absorbed and justified by future expected benefits. Establishments vary in their ability to absorb such costs; however, they are more mobile than transportation systems, which once on the ground become stable elements of urban structure. The huge costs of transportation systems, their traditional linkage to utilities systems, plus the controversial public aspect of decisions for their location etc. greatly contribute to their stability over space and time.

The residential establishment is a highly mobile unit in its response to shifts in employment opportunities; it seeks some value weighted equilibrium between the length and ease of the journey to work[24], and economy in housing expenditure. Carroll has studied six cities with information on the location of employment and residences dating back to the first quarter of the century; he concludes that forces are constantly at work tending to minimize home-work separation.[25]

---

[24]The cost of trip is assumed to be directly proportional to its length.

[25]Carroll, op. cit., the cities are: Detroit-1914, Chicago-1915, Pittsburgh-1917, Washington-1925, Baltimore-1925-26, Milwaukee-1927.

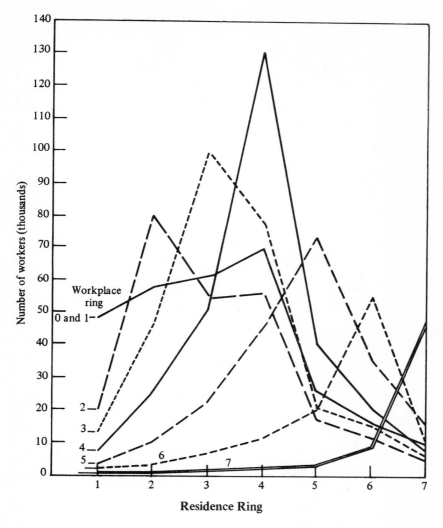

Figure 7. Number of Chicago white workers residing in each ring by workplace ring. Source: Tabulated from Chicago "first work-trip" file.

Kain arrives at similar conclusions based on late 1950's data for Chicago and Detroit. Figures 7 and 8 show that:

> ... with a few exceptions for the innermost rings, the residence ring housing the largest proportion of each work place ring to workers is also the same ring in which they are employed. In

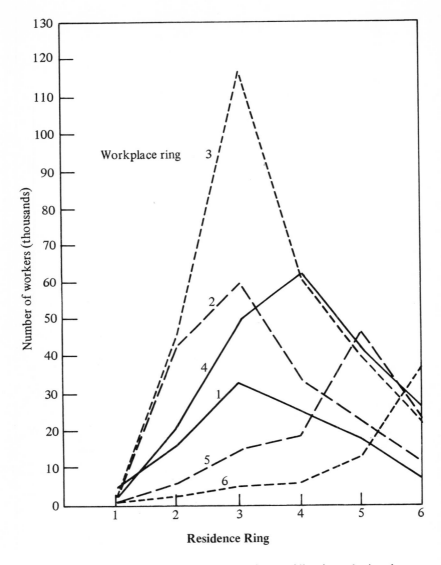

Figure 8. Number of Detroit white workers residing in each ring, by workplace ring. Source: Tabulated from Detroit "first work-trip" file.

Source of Figures 7 and 8: Meyer, John R., F.F. Kain, and M. Wohl, *The Metropolitan Transportation Problem,* Cambridge: Harvard University Press, 1965, pp. 122-123.

short, workers tend to live in their work place rings; furthermore this tendency is greater for outer than inner work place rings. This suggests that a tendency exists to try to economize on transportation outlays for work-trips and that this may be somewhat easier to achieve if one's work place is at a distance from the CBD[26].

Two studies, one in Boston and the other in Washington, D.C., have investigated the extent of residential mobility in response to the relocation of employment centers on the fringes of the urban area. Findings were that between 40-60% of employees moved their homes closer to their place of work; those who did not move either found that the commuting work-trip was too inconvenient for them and quit their jobs or they valued their social ties in their community more than the proximity to work.

It seems appropriate here to introduce some of Kain's findings, relating the effect of the race problem in housing market segregation to the travel behavior of urban residents:

> Centrally employed nonwhites make short work-trips and probably shorter than they would freely choose if the housing market was not segregated. Centrally employed whites seem to do just the opposite. Noncentrally employed nonwhites, seem to travel relatively long distances to work while noncentrally employed whites usually manage to live close to their work. In general, the evidence is that discrimination forces minority groups into a disproportionate amount of cross-hauling and reverse commuting. Ghettos and their counterparts are usually near the CBD; accordingly since more and more work places are located at the fringes of cities, more and more negroes will be traveling to and from work in directions opposite to the main commuter streams unless housing discrimination is lessened.[27]

We can also look to the behavior of individuals between urban activities as generating activity systems. In developing the concept, Chapin[28] has sought a better understanding of the structure and performance of urban communities and movement systems. He clarified the characteristics of interaction which would differentiate between types of activity systems. These characteristics can be summarized in the following:

1. *Spatial boundary of system* — either no specific boundary, e.g., regional, national, or one which has a locus in a particular metro area at a particular time.
2. *Components of interaction* — 'within' interaction or 'between' interaction.

---

[26]Meyer, John R., J. F. Kain, and M. Wohl, *The Metropolitan Transportation Problem*, Cambridge: Harvard University Press, 1965, Chapter 6, pp. 123-124.

[27]Kain, op. cit., Chapter 7, pp. 166

[28]Stuart Chapin, *Land Use Planning*, 2nd edition, Chapter 6.

3. *Dynamics of interaction* — systems that are relatively stable involving slow changes in spatial distribution, or systems which are transitory and bolatile involving rapid change.
4. *Recurrence of interaction* — systems recurring in identifiable spatial problems, or activity systems recurring in random spatial patterns.
5. *Interest served in interaction* — systems which involve the public interest and have a relationship to the physical layout of the community, or interactions between private interests not involving community consideration.

Given this classification, it becomes obvious that the work-trip or journey to work corresponds to an activity system which has a locus in a metro area, which as a 'between interaction,' is relatively stable, recurrs in identifiable spatial patterns, and involves the public interest in its relationship to the physical layout of the community. The above illustrates the possible connection between activity systems generating types of trips and transportation systems serving one or a variety of trip types.

If we consider the previous five groups of system characteristics, we can list thirty-two combinations $(2)^5$ which would possibly identify corresponding types of activity systems and types of trips. If these types could be grouped according to similarities in service requirements and then matched with the performances of movement systems which are most appropriate to satisfy their demand, then the relationship between urban activity and transportation systems would cease to be dependent, in the main, on one mode and one type of channel.[29]

The multiplicity and complementarity of movement systems could result in less concentration and a greater mixture of urban activities and consequently more of the 'within' interaction with less of the 'between' trips. According to the technology used, such movement systems could become less stable and more responsive to changing conditions. Increased flexibility makes it feasible to optimize routing and the location of the facility such that the length of any trip is minimal.

Improving the performance of the transportation network and improving the location of the network relative to air-current patterns are the two areas which remain to be discussed. Our concern here is not to discuss technical aspects of these problems as much as it is to point out some of the implications on the urban spatial structure.

4. *Improving the performance of the transportation network.*

From data such as that collected by Alan Voorhees and the National Air Sampling Network, NCAPC, (Tables 4 and 5), we find that the largest amount of contaminants from the automobile is emitted during the phases of acceleration and deceleration, that fuel consumption increases as average speed is reduced and that contamination is highest in signalized arterials and the

---

[29]Of interest and relevance is Volume II of Stanford Research Project — *Future Transportation Systems and Their Implication on Urban Life and Form*, Stanford Research Institute, Menlo Park, Calif. 1968

Table 4

Percentage of Contaminant Emitted per Mile, in Each Mode of I.C.E. Operation

	Gross Hydrocarbons	Carbon Monoxide	Oxide of Nitrogen
Idle	5.9%	7.5%	0.03%
Cruise	14.1%	14.3%	21.40%
Acceleration	56.2%	62.2%	78.50%
Deceleration	23.7%	16.1%	0.17%

Source: Alan M. Voorhees, *"Variables Affecting Traffic & Vehicular Operating Conditions in Urban Areas;"* January 1968

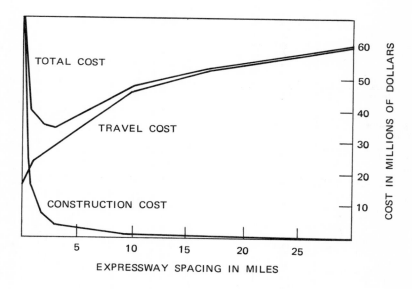

Figure 9. Construction costs, travel costs and total costs as a function of expressway spacing for an average square mile of area.

Calculated for an area of 20,000 daily trip destinations per square mile, with expressway construction costs of 8,000,000 dollars per mile and arterial streets one-half mile apart.

Source: Chicago Area Transportation Study, Volume III.

Table 5. Commuter Exposure to Carbon Monoxide.

20-30 Min. Runs in Traffic
(Concentrations in ppm)

City	Year	Route Type	No. Runs	Exposure During Run			Brief Peak Exposure		
				Min.	Avg.	Max.	Min.	Avg.	Max.
Atlanta	1966	Expressway	10	12	23	35	50	73	92
		Arterial	9	21	30	40	37	62	91
		Center City	7	13	21	27	33	49	75
Baltimore	1966	Expressway	18	2	9	21	6	32	79
		Arterial	25	6	17	33	14	37	82
		Center City	14	15	24	38	28	54	92
Chicago	1966	Expressway	34	13	24	37	24	52	91
	1967	Expressway	34	3	20	35	7	43	86
	1966	Arterial	51	7	16	31	16	32	59
	1967	Arterial	29	1	16	41	3	32	68
	1966	Center City	17	24	32	55	36	58	91
	1967	Center City	10	19	34	53	45	70	109
Cincinnati	1966	Expressway	7	3	10	15	8	22	32
	1967	Expressway	24	7	15	34	12	27	53
	1966	Arterial	25	3	17	52	6	36	79
	1967	Arterial	11	9	15	19	18	32	50
	1966	Center City	8	17	29	50	30	56	96
	1967	Center City	6	12	20	32	19	36	71
Denver	1966	Expressway	24	10	21	38	20	40	71
		Arterial	41	16	32	61	35	65	138
		Center City	10	21	34	54	36	69	96
Detroit	1966	Expressway	48	11	25	54	22	51	116
		Arterial	51	14	25	41	34	54	91
		Center City	16	15	25	36	33	51	85
Houston	1966	Expressway	31	2	16	39	9	29	74
		Arterial	23	5	15	33	13	36	82
		Center City	14	16	38	70	32	75	98
Los Angeles	1966	Expressway	87	9	29	62	23	52	102
		Arterial	15	24	38	60	45	74	147
		Center City	17	27	40	60	44	68	111
Louisville	1966	Expressway	14	2	8	21	6	17	30
		Arterial	14	4	16	31	8	32	61
		Center City	10	16	23	33	28	40	62
Minneapolis-St. Paul	1966	Arterial	59	10	23	41	18	52	130
	(M.)	Center City	15	20	30	45	35	52	>91
	(St.P)	Center City	13	21	35	65	28	56	>91
New York	1966	Expressway	31	8	23	44	15	47	91
	1967	Expressway	17	5	21	39	16	45	141
	1966	Center City	47	14	32	58	24	57	91
	1967	Center City	30	9	27	42	16	50	119
Philadelphia	1967	Expressway	20	8	17	38	17	38	94
		Arterial	40	9	22	38	17	40	82
		Center City	20	18	29	38	37	55	80
Phoenix	1966	Expressway	39	12	23	50	18	40	83
		Arterial	26	13	27	43	26	56	87
		Center City	12	27	38	54	44	72	105
St. Louis	1967	Expressway	37	2	10	22	7	22	57
		Arterial	38	8	17	34	17	36	70
		Center City	19	15	28	50	28	56	109
Washington	1967	Expressway	35	2	12	33	5	27	76
		Arterial	17	10	20	30	22	40	86
		Center City	34	8	26	63	13	45	116

Source: National Air Sampling Network, NCAPC.

158

local streets of the Center City. This information means that the amounts of emitted contaminants seem to be directly proportional to the frequency of vehicular STOP — GO operations. This suggests that one way to reduce contaminant emission would be to increase uninterrupted vehicular flows at higher speeds by minimizing the number of intersections per lineal mile of street network.

Creighton, et al[30], as well as the Chicago Area Transportation Study, have shown that a relationship exists between the cost of the network and the spacing of arterials and expressways, (Figure 9).

There is an optimum spacing where the combined costs of movement and construction are minimized. The spacing of arterials is a function of population density, car ownership, capacity, and average speed of network. Hence, if appropriate models are developed, we can either determine (given some tolerable level of contamination) the characteristics of the network and its cost per mile or determine the dollar costs and characteristics of the network. We can then evaluate whether or not the resultant contamination is within acceptable limits.

The reasonableness of minimizing network intersections would depend on the following factors:

a. The spatial pattern of community activities and our ability to effectively separate pedestrian and vehicular traffic. Even then we cannot eliminate the points of contact between the two, especially where the performance of vehicles and of network is scaled down to the level and pace of the pedestrian.

b. Ability to separate the traffic flows in different directions by using multi-level networks, overpasses or underpasses. Although this could be easier in new networks than in existing areas, the cost of multi-levels and the resulting complexity at points of interchange may make large scale application unfeasible.

c. Ability to reduce congestion and increase efficient use of the network capacity by a variety of ways such as staggering work hours to reduce peak hour loading of vehicles and establishing through legislation a standard for 'vehicle space per user' where any excess of that standard would be subject to taxation.

d. If evenflow movement can be achieved between nodes of activity, the difficulties of achieving it within a node are apparent due to the fragmented distribution of activities, the nature of pedestrian movement and the desire for the convenience of close access to destination. Hence, the reasonableness of an evenflow network would depend also on our ability to incorporate within a node, mechanical

---

[30] For a full discussion of the concept and derivation of mathematical equations see: Creighton, R. L. Hock & Schneider "Estimating Efficient Spacings for Arterials and Expressways", paper presented before Highway Research Board, 39th Annual Meeting, Washington, D.C., Jan. 1960, published in HRB Bulletin 253.

non-polluting movement networks, which in maintaining a high level of safety, comfort and convenience, would become the link between the pedestrian, his vehicle and his activities. Such mechanical systems are technologically feasible and operational.

Finally, we should not forget that the quantity of emitted contaminants in any period of time is a direct function of the number of cars in use multiplied by the average emission/car in that period and that the emission/car is a function of the performance of the engine as well as the performance of the network on which the engine operates. Hence, the total amount of contaminants would decrease only if: a) both the number of cars in use and the average emissions per car decrease; b) if one of them remains constant while the other decreases. But we know that the number of cars in use is increasing every year in response to increase in population and increasing car ownership; hence, vehicular contamination could not be lessened unless the average contaminant - emission/car is decreased at a rate faster than the increasing rate of cars in use.

5. *Improving the relation between the location of urban development and transportation network and the meteorology of the area.*

Although a profusion of literature[31] has dealt with the fixed source of pollution and its relation to meteorological conditions as a determinant of its location, very little is available on the spatial relation of major traffic flows to air current patterns. One of the few is Alexander and Manheim's experimental work[32] where air pollution patterns were one of the 26 variables used to determine the location of a highway route.

The literature annotated by Pelle shows the extent to which the physical development of urban areas modifies the climate of the environment; many articles or books propose guidelines for using geography, climatology and meteorology as determining factors in the physical planning and designs of urban communities. The primary concern in these guidelines has been to minimize air pollution from stationary sources; however, one can detect that the reduction of vehicular contamination has been of secondary concern, seemingly based on the assumption that it could be achieved through better diffusion in air currents rather than the control of emission at the vehicular source.

If anybody wonders why these guidelines or controls are still in the realm of 'theory,' one can only speculate that the application of guidelines or controls will either increase the cost of urban development or reduce the opportunity for profit. It seems that our social values are just beginning to get concerned with controlling air pollution. We have yet to see the long term

---

[31] See William J. Pelle, *Annotated Bibliography on the Urban Planning Aspects of Air Pollution Control*, for Northeastern Illinois Planning Commission, and United States Public Health Service, March 15, 1966.

[32] Christopher Alexander and Marvin Mannheim: *Use of Diagrams in Highway Route Location: An Experiment*, Cambridge, MIT, School of Engineering, 1962, Pub. No. 161.

benefits of investing in environmental quality, to accept paying the extra cost of development or to forego some immediate profits.

Another one of the difficulties in the control of vehicular air pollution is that although progress is being made in simulating and explaining the nature, concentrations, and configurations of pollutants at the exhaust source and at their concentrations in the smog cloud, great difficulties still exist in identifying the factors and explaining the dynamics of dispersion, movement and accumulation of vehicular contaminants. The role of natural and induced air currents, especially at the regional scale, in such dynamics is not yet understood, although the beneficial effects of westerly winds in occasionally clearing the smog cloud over Los Angeles can readily be observed and appreciated.

I have attempted to develop a simple typology matrix depicting areas of maximum and minimum air pollution as an interaction between a range of wind directions and gross topographical configurations. Due to the many meteorologic, climatic and geographic variables involved, this attempt was deemed futile by geographers and meteorologists who were consulted; they suggested that except for very gross and rather useless generalizations, each case and each location should be considered on its own if any operational determinants are to be derived.

IV. SOME POLICY QUESTIONS

In the course of discussing some relationships which may affect the control of vehicular pollution in urban areas. some policy questions are implicit in the discussion. Although urban planners have no ready answers for them, I believe they should be considered by those concerned with the quality of air in our environment. I hope that the dialogue will point to some answers which will help make the control of vehicular contamination operational.

1. We have assumed that in the foreseeable future the vehicular source of air pollution (the internal combustion engine) will remain unchanged in its operation characteristics. Is this a valid assumption? Present developments in research indicate that a high degree of control of that source is technologically feasible, albeit with some upper limit; however, is this control imminent? The delay seems to be couched in the conflict between public and private interests for desirable air quality versus economic costs to industry. If resolved, the problem may be reduced in magnitude to one of controlling pollution from stationary sources; this control is progressing at present in large metropolitan areas with varying degrees of success and may be not as rapid as desired.

2. We are dealing with an existing urban environment where huge public and private investments have already been sunk in existing conditions. Is the huge cost of reducing and/or eliminating vehicular contamination by changing the urban environment justified by the benefits of reducing and/or eliminating the effects of that contamination? What are the dimensions of the problem? When does it become critical rather than a nuisance? What are acceptable levels of human tolerance and when does exceeding them become critical? Is

there a threshold beyond which air can never be decontaminated and where, possibly, human beings adapt to that air quality as being 'clean-standard' and with no ill effects?

3. In the control of vehicular air pollution, one can only assume and expect that resources will be limited and that some tradeoffs will have to be made. Is there an optimum allocation for intervention and control efforts between controlling the source and controlling the environment to achieve some desired level of pollution-free air? If controls are developed (for the various system variables affecting vehicular air pollution) and implemented, how much would each controlled variable contribute toward decreasing pollution and at what cost? Hence, where should the real intervention effort be applied? Is there a proper sequence of research and development and implementation applicable to each system variable? Where is the highest payoff?

4. In the increasing mobility of people there seems to be a self adjustment, over time, to some average distance from work. As vehicular work trips account for about 60% of all trips, a policy could be to change the spatial distribution of activities and/or combination of modes so that this average work trip length is reduced. What is the cost of this reduction? Is it justified by benefits of reducing contamination? Who makes the tradeoff choice between comfort in the reduction of air pollution versus the comfort of mobility, convenience, status, etc.? Variations in trip length and volume/purpose of trip resulting from combinations of alternative transport modes are not always large enough to make a clear cut choice for the decision-maker.

5. Unless some strict social and economic controls are imposed, it is impossible to enforce strict proximity of residence to work as well as centralization and compactness of urban activities. The nature of our pluralistic society and our objectives of maximizing freedom of choice and the opportunity to exercise a wide range of preferences work against such an imposition.

6. Generally speaking, we are dealing with a multiple objective and multiple means situations. It is, therefore, necessary to develop a systems evaluation of the relationships between objectives and the different means; it is important to know the impact of intervening with one objective on the achievement of other objectives.

Comments Relating To: *Vehicular Air Pollution:*
Variables Influencing The Urban Transportation System
by I. M. Jammal

By G. A. Fleischer

Dr. Jammal's paper is clearly written, direct, and, we would suppose, a reasonably comprehensive discussion of the current state of the art of urban transportation as viewed by the city/regional planner. The paper is not intended to serve as an encyclopedic treatise; indeed, we are not competent to judge the extent to which the author has surveyed the field, although we are satisfied that the breadth of coverage is impressive.

In general, the paper serves admirably to educate the reader about the various extant theories of urban structure and growth . . . and the somewhat more limited discussion of the relationships between urban form and transportation needs. We are somewhat disappointed, however, that the practical implications are not explored more fully. For example, recent research findings indicate the relationship between urban form and trip length, and new communities could be designed to control this characteristic. But what can be said about existing communities? Clearly, our focus here must be upon the cost of bringing about changes in urban form and the net benefits accrued as the result of these changes. (The author, by the way, is not unaware of the importance of this issue: cf. p.152.)

Dr. Jammal asserts (p.126) that "fixed sources of pollution are responsible for 39% of the total emission, while the ICE, internal combustion engine, in transportation is responsible for the largest share of contamination namely 61% (sic)". This is a statement of gross statistics, of course, and deviations from these mean values are substantial and significant. In the Los Angeles basin for example, stationary sources currently account for only about 10% of contaminants. Moreover, the "contributions" of different sources to air pollution should be associated with various types of emittants, e.g., the ICE contributes much of the CO in the area but relatively little SO_2. The importance of the urban transportation system to the "air pollution problem" is therefore largely a function of the particular community under discussion and the type(s) of pollutant(s) of interest.

In response to these and other comments, we would hope that Dr. Jammal will respond to the following question: Does the transportation system follow from the spatial distribution and activities in urban areas, or do form and activities follow from the transportation system(s) available to the community? These are inter-related issues, of course, but what are their relative contributions?

A statistical analysis of the Voorhees data included in Table 3 (p.148) and plotted in Figures 6 A, B, C (pp.149-51) would be helpful. It is not at all clear that "length of trip (in miles and time) increases with the size of the urban area, but at a decreasing rate," nor is it equally clear that the average speed of the network is independent of the increase in the size of the urban area. Moreover, and perhaps more important, statistical correlation or curve fitting tells us nothing and suggests little about the causal relationships at work here.

The annotated bibliography of Pelle is of very great significance to the objectives of this paper. In particular, we are interested in Dr. Jammal's observation (stemming from this bibliography) that "the reduction of vehicular contamination is of secondary concern and that it could be achieved through better diffusion in air currents rather than the control of the emission source by any number of means." What are the experimental evidence, if any, which support this statement? Is it clear that the latter class of countermeasures is superior to the former? And by what criterion measure is this so?

The policy questions outlined by Dr. Jammal in the concluding section of the paper are especially valuable as they relate to our efforts here. It would seem to us that these six or seven questions could well serve as the basis for a summary document for the symposium.

Discussion of Paper, "Vehicular Air Pollution:
Variables Influencing The Urban Transportation System"
by Ibrahim M. Jammal

By Gerald Fleischer, Ph.D., P.E.

Associate Professor
University of Southern California
Department of Industrial and Systems Engineering

and

Akira Hasegawa
Regional Planner
Los Angeles County Regional Planning Commission

Comments by Dr. Jammal

The City Planner is one of the many specialists who are concerned with the quality of the environment and its effects on the physical and mental well being of people living in it. The planner may be able to manipulate those urban, spatial, organization variables which affect air pollution. And, in contributing to the solution of the problem, the city planner needs to exchange and share relevant information with pollution scientists and control officials.

Pollution is still low among the priorities of planners and citizens; economic considerations of urban development and growth take precedence. The reduction or elimination of air pollution should be an important criterion in city planning and high on the list of citizens concerns.

Scientists and pollution specialists should communicate their research findings on the nature and deleterious effects of air pollution in an understandable format which is meaningful for lay citizens and other professionals concerned with the problem. This kind of communication would lead to intelligent citizen efforts for political action and operational guidelines for control officials. We should realize that the quality of air is only one of several variables which influence and describe the total quality of the environment. If we have a better understanding of the proper role and importance of air quality relative to these other variables, we would be able to expend our efforts and resources more efficiently in dealing with the total pollution problem.

There are several aspects of urban structure and spatial organization which could influence the quantity of vehicular contaminants emmitted in urban areas. These are the relationship of trip length to the shape of the city and to its size; the relationship between the transportation network, the performance of the network, and the location of urban activities; and, finally, the relationships between the location of urban development, the transportation network and the meteorology of the area. This last relationship has been the least researched and the least considered in city planning. Realistically, these relationships are more apt to be implemented in new cities than in existing ones.

A possible way to control the number of daily work trips and the length of those trips would be to induce a shift in the location of workers closer to their place of employment. One strategy might be to make the residential property tax a function of the distance between the residence and place of employment; such a tax could be further modified with respect to the mode of transport used between home and place of work.

Unless the benefits are clear cut, the cost of altering existing conditions will seem high to those interests which have already committed huge capital investments. We cannot keep talking about the city as if it is some entity which just materializes at some point in time. We need to keep in mind that the city grows and changes over time in response to the interplay between prevailing institutions, people's decisions and a variety of social, political and economic forces. In turn, one existing physical manifestation of the city becomes quasi-permanent and a strong force in influencing the course of subsequent development or changes.

Dr. Gerald Fleischer:
Although a wide range of theories pertaining to urban structure were explored, a fuller exploration of the relationships between urban form and transportation needs and their practical implications on the problem of air pollution would be useful. Which comes first, urban activities or the transportation network? Inasmuch as they are interrelated, what are their relative contributions to the problem discussed here?

Something can be done concerning the form of new cities or growing cities, but what should be done about the existing cities and their older core areas? A fuller statistical analysis of the Voorhees data is needed to validate the conclusions derived from the scatter diagrams as well as more impirical evidence to compare alternative ways of reducing pollution by either controlling the source or reorganizing the environment.

Akira Hasegawa:
A planner's job is very difficult in that he has to contend with a multiplicity of jurisdictions and political boundaries which overlap and intersect in the region. For example, Los Angeles County has over seventy incorporated cities each of which has its own planning and zoning activities. Things would be much easier if the planning and zoning were done on a metropolitan basis; many conflicts between the goals and objectives of each jurisdiction would be eliminated.

Another problem is the rapid pace of change where a decision today may not be applicable tomorrow. The general unwillingness to compromise among public officials, planners, residents and developers on issues of land development usually ends in litigation or the acceptance of decisions inequitable to the residents. It is necessary and important to make tradeoffs among the above factors in setting goals and means of implementation.

166

Even if another automobile engine is developed, nitrogen oxides will still be produced as long as there is any type of combustion. Daily automobile trips, other than home-to-work, have to be considered, but this depends on the nature of the urban area under study. In Los Angeles, people are in love with their cars; they value too highly the individual home style of life and the amenities of outlying areas to either opt for denser development or seek other means of transportation.

We may have to have a crisis before anything is really accomplished in controlling air pollution. Meanwhile, we should make full use of zoning, taxation and planning to influence the use of land and the location of urban activities with respect to the transportation network. Fourteen years ago, zoning, taxation and planning were not really considered in the control of air pollution; it is becoming apparent that they could become very effective methods if the difficulties with political jurisdictions and public cooperation could be overcome.

Professor Jammal:

The conclusions drawn from the scatter diagrams in the text are premature and need a fuller statistical analysis. As to whether urban activities came before transportation or vice-versa, the two are so inter-related that they have to be considered simultaneously. In general, that question is similar to whether the chicken or the egg came first. The question could be answered partially if we consider the transportation system as a series of components fulfilling different functions in meeting transport demand. We may establish the sequence in which the components are built. We may then evaluate how the sequence responds to transport demand generated by urban activities as well as types of activity which are located in response to the accessibility generated by that component of the transportation system. There is a need for pollution researchers to become concerned with studying the effect of pollution on the urban environment in which it is generated, as well as its effect on the behavior of the people in that environment. In reviewing a very recent federal publication on abstracts of air pollution studies, out of 1000 entries, about one percent of the studies were concerned with air pollution in the urban environment.

Dr. Morris Neiburger, Professor of Meteorology at UCLA:

Good city planning can help disperse air pollution by laying streets along breeze lines, thus causing the contaminants accumulating along the axis to disperse more rapidly. High and low rise buildings can be alternated to cause air turbulence and trees may be provided in green belts to absorb some of the pollution.

Professor Jammal:

First, air pollution should not be controlled by air currents. They should be controlled at the source. Second, the issue is not whether air pollution should be controlled, but rather which method would achieve the objective more effectively, controlling the source of controlling the environment.

Dr. John R. Goldsmith, M.D. — State Department of Public Health

Both aspects are important and more research has to be done in learning about the capacity of greenbelts and plants, in general, to absorb pollution.

The application of cost-benefit evaluation is defeatist and we should think of builidng new cities and new environments; we should have the legal power to stop growth of existing cities until we can plan properly and resolve major environmental problems.

Dr. Arthur Atkisson, Co-Chairman of the Symposium:

A recent poll indicated that the residents of Los Angeles area still wanted growth to continue, wanted to live in a single family dwelling with a view and a two or three car garage, as well as an end to air pollution. With these desires in mind, even if air pollution abatement could be achieved in another ten years then it will still be too long a period. Air quality standards would have to be carefully phrased so as to pursuade and convince everyone of the need for action.

MATHEMATICAL MODELS OF AIR QUALITY
CONTROL REGIONS

By John H. Seinfeld
Chemical Engineering Laboratory
California Institute of Technology
Pasadena, California 91109

Introduction

The control of air pollution is one of the most important problems facing the large urban and industrial centers of the world. Because of the multitude of factors which influence the existence of an air pollution problem, a systematic approach to the control of air pollution is required. Perhaps the first step is the delineation of geographical regions which have a common air pollution problem. This is being done, resulting in the air quality control regions in accordance with the Clean Air Act of 1967. The control of air pollution in each region can be considered a separate problem, even though different regions may require similar control measures. The regions are being established primarily on the basis of meteorological and topographical considerations and may be visualized as huge three-dimensional atmospheric volumes in which air pollution is to be controlled.

In order to set standards for sources of airborne pollutants, criteria based on the effects of air pollution on receptors, e.g. human health, plant life, materials, visibility, etc., must be determined. For example, the California Department of Public Health has established the following standards for carbon monoxide: 30 ppm for an 8-hour exposure; 120 ppm for a 1-hour exposure. Standards for nitrogen oxides have been based on atmospheric coloration and levels likely to cause long term health impairments. Oxidate levels — mainly ozone — have been tentatively set as 0.1 ppm for a 1-hour average, not to be exceeded more than one percent of the time.

Since it is highly unlikely that air pollution control will be achieved by modification of the state of the atmosphere, the control of air pollution must be directed at the sources of pollutants. For example, the 1971 California standards for automotive exhaust emissions are 2.2 grams per mile of hydrocarbons, 23 grams per mile of carbon monoxide and 4.0 grams per mile of nitrogen oxides.

The question then arises: Will the source emissiom standards that are enacted be consistent with the air quality standards in a particular air quality control region? In other words, knowing the amount, location, and composition of the inputs, the source emissions, can the level of the result, the concentration of air pollutants, be determined at any location in the air quality control region? Conversely, if we set certain standards for air quality, can we determine the inputs necessary to meet these standards? These critical questions can only be answered by means of mathematical models for air quality control regions. Current air quality standards do not delineate pollution levels at different locations in an urban area. It is conceivable that a particular area in a city could experience levels exceeding overall air quality standards while the mixed average levels over the entire area would be well below the standards. It is thus necessary to have reliable models for predicting pollutant levels at any location in an air quality control region given source data and meteorological conditions.

Mathematical models of air quality control regions will facilitate the determination of the degree to which each source contributes to the overall pollution problem. In addition, planning of the location of future sources so that air pollution is minimized can become a factor in urban planning. From a model one would be able to predict the meteorological conditions under which a region might expect abnormally high air pollutant concentrations and provide a warning for reduction in emissions during such a period.

For our purposes air pollution will be classified according to two categories: non-chemically reacting, or inert, and chemically reacting. The constituents emitted directly from sources are usually termed primary contaminants. If these primary contaminants are mainly responsible for the observed effects of the air pollution, i.e. they are dispersed in the atmosphere but do not undergo chemical change in the atmosphere, then these air pollutants will be called non-chemically reacting. Typical examples on non-chemically reacting air pollutants are sulfur dioxide,† carbon monoxide and solid particulate matter.

Constituents which are not emitted directly from sources in appreciable quantities but are formed in the atmosphere from chemical reactions among the primary contaminants are called secondary contaminants. If the secondary contaminants are mainly responsible for the effects of the air pollution then the air pollution is of the chemically reacting type. The most prominent example of chemically reacting air pollution is photochemical smog, in which inputs of nitric oxide and hydrocarbon vapors are transformed by atmospheric chemical reactions into ozone, nitrogen dioxide, and oxidized organic products. Although the same general principles will apply in modeling non-chemically reacting and chemically reacting air pollution, the latter introduces much more complexity into the problem.

The mathematical modeling of an air quality control region for chemically reacting air pollution includes the following steps:
1. Determine the meteorological factors influencing the spread of pollutants.
2. Determine a suitable kinetic mechanism for the atmospheric chemical reactions.
3. Combine these two to formulate a model predicting pollutant concentrations as a function of time and location.

The objective of this report is to set out the ground rules for the mathematical modeling of polluted atmospheres. Our ultimate interest is the chemically reacting situation, and, in particular, photochemical smog, although everything that is said will apply to the non-chemically reacting case with appropriate simplifications. First, the effects of meteorology on air pollution will be discussed. Then, a brief discussion of the chemical reactions of photochemical smog will be presented. Finally, various techniques for the mathematical modeling of inert and chemically reacting air pollution will be reviewed and examined.

Meteorological Effects on Air Pollution

In this section the effect of meteorology on air pollution is outlined. Much of the discussion has been adapted from the review of Panofsky (1969) on this subject. We will concentrate on the role that each of the factors plays in the development of an atmospheric model.

Panofsky (1969) has divided the effect of the atmosphere on pollutant concentrations into three parts:
1. The effect on the "effective" emission height.
2. The effect on transport of the pollutants.
3. The effect on dispersion of the pollutants.

In addition, it is useful to add a fourth catagory·
4. The effect on the chemical reaction rate constants.

The effective height of a source is normally used for calculation of transport and disperson and is assumed to be slightly upwind of a point straight above the stack. The distance above the stack is determined by the amount of rise of the effluent from the source. The amount of rise depends on meteorological and nonmeteorological factors. The meteorological variables affecting plume rise are the wind speed, the lapse rate (the decrease in temperature with height) and the difference between the temperature of the effluent and the temperature of the air. The nonmeteorological variables affecting plume rise are the source area and the efflux velocity. There are a large number of formulae which relate plume rise to the meteorological and nonmeteorological variables. Briggs (1968) has summarized the most commonly used rise equations.

The transport of airborne pollutants is due to the wind. The concentrations of pollutants at different locations in a region will be strongly dependent on the mesoscale wind field, the wind patterns on the scale of a few miles to one hundred miles. Microscale wind patterns (on a scale of a few hundred feet) are probably more important in dispersion than in transport.

The dispersion of a pollutant depends on the mean wind speed and the characteristics of the atmospheric turbulence. Atmospheric turbulence is composed of horizontal and vertical eddy motion which tends to mix packets of air. Atmospheric eddies are formed by free convection and wind shear. Free convection occurs whenever the temperature decreases rapidly with height influencing circulation of air because of buoyancy effects. Wind shear induces mechanical turbulence close to the ground. This turbulence increases as the wind speed increases and is greater the rougher the terrain. The surface roughness is usually characterized by a roughness length z_0 which is proportional to the size of eddies that can exist among the roughness elements.

Dispersion is also ultimately influenced by the existence of a temperature inversion. Normally, the temperature of the atmosphere decreases with height, however, there often exist small layers in which the temperature increases with height. Such layers, called inversion layers, inhibit mixing between the layers above and below. Warm air on top of cold is stable, and free convection will not take place readily. Thus, the inversion acts as a lid on the atmosphere, containing pollutants in a fixed atmospheric volume. The existence of a temperature inversion at 1000-1500 feet is one of the key factors in the Los Angeles air pollution problem.

The chemical reaction rate constants are influenced by temperature for thermochemical reactions and sunlight intensity for photochemical reactions. The rates of thermochemical reactions depend exponentially on temperature, in some cases approximately doubling for each $10°$ rise in temperature. Solar radiation intensity is an important variable in the photochemical smog reactions, wherein a primary step is the absorption in the ultraviolet region by NO_2 and organic nitrites.

Photochemical Smog

The most well-known form of chemically reacting air pollution is photochemical smog. Photochemical smog is produced by reactions involving NO, NO_2 unburned hydrocarbons and oxygen. The smog formation process is characterized by the oxidation of NO to NO_2, the oxidation of unsaturated and aromatic hydrocarbons to aldehydes and ketones, and the formation of O_3 and peroxyacyl nitrates (PAN's). In this section the general nature of the chemical reactions of photochemical smog will be outlined.

The high concentrations of O_3 found in photochemical smog cannot be explained by direct light absorption by O_2. The short ultraviolet wavelengths necessary for this reaction, while prevalent in the upper atmosphere, do not reach the lower levels of the atmosphere. Also, the high concentrations of NO_2 attained from an input of NO cannot be explained by the thermal reaction of NO and O_2 which proceeds far too slowly to be of any importance in smog formation. A photochemical process is thus apparently intimately involved in smog formation and consideration of the components of a polluted atmosphere isolates NO_2 as the only photochemically efficient absorber of the ultraviolet and visible light reaching the earth's surface. It has now been firmly established that photodissociation of NO_2 from absorption of sunlight in the range 300-4000 Å is the primary photochemical process in the smog reactions.

$$NO_2 + h\nu \rightarrow NO + O$$

The reactive oxygen atoms react quickly with O_2 to form O_3

172

$$O + O_2 + M \rightarrow O_3 + M$$

where M is a third body (any other molecule in the system). Initially O and O_2 collide to form an energetic O_3 molecule. Then collision of this O_3 with another molecule in the system enables a transfer of some of the excess energy to the other molecule to produce a stable O_3 molecule. Even with the requirement of a triple collision this reaction is extremely fast. NO and O_3 react rapidly to regenerate NO_2 and O_2,

$$NO + O_3 \rightarrow NO_2 + O_2$$

Each of these reactions is extremely fast, at least an order of magnitude faster than other reactions occurring in the atmosphere. There are several other reactions involving NO, NO_2, O, O_2, O_3 and other oxides of nitrogen but these three are generally accepted to be the most important.

Stephens (1968) has calculated the O_3 concentration which might be formed at steady state by irradiating any mixture of NO, O_3, NO_2 and O_2 in the absence of other species. He estimates that 10 pphm of NO_2 in air would yield 2.7 pphm of O_3, while 100 pphm of NO_2 would yield 9.5 pphm of O_3. This calculation was based on zero initial NO, and these O_3 concentrations are still far below those sometimes found in polluted urban air. In fact, most of the oxides of nitrogen are emitted as NO, which would prevent any appreciable O_3 concentration from being reached as a result of these reactions alone. Thus, other reactions must be contributing to the smog formation process.

A process which is slow compared to these reactions but which converts NO to NO_2 without consuming an equivalent amount of O_3 will lead to the accumulation of O_3. The presence of reactive hydrocarbons provides such a process. Both O and O_3 have the ability to react with and oxidize hydrocarbons. The intermediates formed in such a reaction are highly reactive and readily participate in further reactions or decompose to yield free radicals. Free radicals, from the reaction of hydrocarbons with O and O_3 and, in smaller quantities, from the photodissociation of aldehydes, ketones and alkyl nitrates, are necessary intermediates in the further reactions of the photochemical smog system.

The free radicals formed can react with O_2, NO, NO_2, hydrocarbons and other free radicals. Reactions with O_2 tend to add the oxygen to the radical to produce peroxy radicals. Reactions of the oxygen-containing free radicals with NO yield NO_2. Reactions of the free radicals with NO_2 generally yield organic nitrates. Reactions with hydrocarbons lead to larger radicals. Evidence of the reaction of hydrocarbons with O and O_3 is based on the abundance of oxygen-containing products which are observed in laboratory simulations and in polluted air. Organic nitrite and nitrate products, e.g. PAN, provide evidence for the existence of free radicals in the system.

Thus, the free radicals participate in a chain reaction mechanism, initiated by the reaction of hydrocarbons with O and O_3, propagated by the reactions with O_2, NO, and fresh hydrocarbons, and terminated by reactions with NO_2 and other radicals. The key aspect of the chain reaction process is that one free radical formed, for example, from the reaction of O and a hydrocarbon will participate in many propagation steps before extinction. A typical history of one such radical might be reaction with NO to give NO_2 and another radical. This radical then combines with O_2 to replenish the oxygen lost to NO. Then the radical participates in a reaction with another NO molecule to generate NO_2, etc., resulting in many molecules of NO converted to NO_2 for each free radical formed. It is this process involving oxygen-containing free radicals which provides the alternate path for oxidation of NO to NO_2 and subsequent accumulation of O_3. The termination steps in the chain reaction explain the existence of many of the oxygen-containing organic products found in polluted air. A highly simplified mechanism incorporating the above aspects has been proposed by Friedlander and Seinfeld (1969),

$$NO_2 + h\nu \xrightarrow{\ 1\ } NO + O$$

$$O + O_2 + M \xrightarrow{\ 2\ } O_3 + M$$

$$O_3 + NO \xrightarrow{\ 3\ } NO_2 + O_2$$

$$OL + O \xrightarrow{\ 4\ } RO_X \cdot$$

$$OL + O_3 \xrightarrow{\ 5\ } PRODUCTS$$

$$NO + RO_X \cdot \xrightarrow{\ 6\ } NO_2 + RO_{X-1} \cdot$$

$$NO_2 + RO_X \cdot \xrightarrow{\ 7\ } PRODUCTS$$

The principal initiation reaction is step 4 (OL denotes an olefin molecule). Many propagation steps are embodied in reaction 6, and termination occurs primarily in step 7.

A pseudo-steady state assumption for O and O_3 yields

$$[O] = \frac{k_1 [NO_2]}{k_2 + k_4 [OL]}$$

$$[O_3] = \frac{k_2 [O]}{k_3 [NO] + k_5 [OL]}$$

If we make the further assumptions that $k_2 \gg k_4 [OL]$ and $k_3 [NO] \gg k_5 [OL]$, then

174

$$[O] = \frac{k_1 [NO_2]}{k_2} = \gamma [NO_2]$$

$$[O_3] = \frac{k_1 [NO_2]}{k_3 [NO]} = \beta \frac{[NO_2]}{[NO]}$$

Finally, if it is assumed that all the free radicals are in a pseudo-steady state, with ϵ the number of radicals generated as a result of propagation and branching in step 6,

$$[R\cdot] = \frac{k_4 [O] [OL]}{k_7 [NO_2] - (\epsilon-1)k_6 [NO]}$$

If $k_4/(k_1 [NO_2] - (\epsilon-1)k_6 [NO])$ is approximately constant and denoted by k_4', the free radical concentration is $k_4' [OL] [O]$.

The dynamic equations for a constant volume batch reactor are thus

$$\frac{d[NO_2]}{dt} = [NO_2] [OL] \left\{ \alpha [NO] - \lambda [NO_2] \right\} \tag{1}$$

$$\frac{d[NO]}{dt} = -\alpha [NO_2] [NO] [OL] \tag{2}$$

$$\frac{d[OL]}{dt} = -[NO_2] [OL] \left\{ \theta + \mu/[NO] \right\} \tag{3}$$

$$\alpha = \gamma k_6 k_4' \qquad \lambda = \gamma k_7 k_4'$$
$$\theta = \gamma k_4 \qquad \mu = \beta k_5$$

Integrated concentration-time curves for this mechanism exhibit roughly the same qualitative behavior as the concentration histories in irradiated dilute automobile exhaust. However, this mechanism has several deficiencies, one of which is the inability to predict as rapid a decrease in NO_2 after its maximum as is observed in actual experiments.

Other kinetic models have been proposed by Saltzman (1958), Leighton (1961), Wayne (1962), and Stephens (1968). Recent work by Wayne and Earnest (1969) has shown that it is possible to model adequately the photochemical oxidation of propylene and nitrogen oxides. It appears that kinetic models that simulate the actual transient behavior of an irradiated air-automobile exhaust mixture will soon be available.

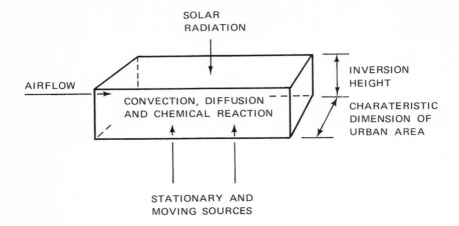

SOLAR
RADIATION

AIRFLOW

INVERSION
HEIGHT

CONVECTION, DIFFUSION
AND CHEMICAL REACTION

CHARATERISTIC
DIMENSION OF
URBAN AREA

STATIONARY AND
MOVING SOURCES

THE ATMOSPHERE AS A GIANT CHEMICAL REACTOR

Figure 1.

Atmospheric Diffusion of Pollutants

For the purpose of mathematical modeling an air quality control region can be viewed as a giant chemical reactor (Figure 1). The primary question that a mathematical model must answer is: given the source locations and strengths and the meteorological conditions, what will be the concentrations of pollutants at any location in the air quality control region? In principle, this question can be answered through solution of the equation of continuity for each species in a turbulent medium (Bird, *et al.,* 1960),

$$\frac{\partial \bar{c}_i}{\partial t} + \nabla \cdot \bar{u} \bar{c}_i + \nabla \cdot \overline{u' c_i'} = D_{ij} \nabla^2 \bar{c}_i + R_i(\bar{c}_1 + c_1', \ldots, \bar{c}_N + c_N') \quad (4)$$

where $\bar{u}$ is the time-average wind velocity vector, c_i is the time-smoothed concentration of species i per unit volume, c_i' and u' the turbulent concentration and velocity fluctuations, D_{ij} the molecular diffusivity and R_i the rate of production of component i by chemical reaction.

It is customary to replace the turbulent mass flux vector $u'c_i'$ by a relationship analgous to Fick's law of diffusion,

$$\overline{u_j' c_i'} = -K_j \frac{d\bar{c}_i}{dx_j} \qquad j = 1,2,3 \quad (5)$$

where the K_j are turbulent eddy diffusion coefficients. Neglecting molecular diffusion relative to turbulent diffusion, we obtain the customary form of the equation of continuity for a species i in turbulent atmospheric transport,

$$\frac{\partial c_i}{\partial t} + \bar{u}(z) \cdot \nabla c_i = \frac{\partial}{\partial x} K_x \frac{\partial c_i}{\partial x} + \frac{\partial}{\partial y} K_y \frac{\partial c_i}{\partial y} +$$

$$\frac{\partial}{\partial z} K_z \frac{\partial c_i}{\partial z} + R_i(c_1, \ldots, c_N) \qquad (6)$$

where the overbar on c_i has been dropped for convenience and the fluctuating terms in R_i have been neglected. z is taken to be the height above the ground.

Most of the efforts to predict atmospheric concentrations of pollutants have centered on inert contaminants, in which case $R_i(c_1, \ldots, c_N) = 0$. Since $R_i(c_1, \ldots, c_N)$ is usually a nonlinear term, the absence of chemical reaction reduces (6) to a linear equation. Numerous solutions to (6) have been obtained for various sources and assumptions on K_x, K_y, and K_z (Pasquill, 1962).

The most serious problem in the exact solution of (6) for inert contaminants is adequate representation of the diffusion coefficients, K_x, K_y and K_z. Since atmospheric dispersion is produced by eddy motion (molecular diffusion is negligible by comparison), the K's are turbulent eddy diffusion coefficients. These coefficients are representative of the product of eddy size and eddy velocity. Both eddy size and velocity depend on many variables, so that the K's are complicated functions of position in the turbulent field. For this reason, exact solutions of (6) have not been highly successful in predicting atmospheric concentrations except in some very idealized cases.

A popular approach, other than solution of (6), is to assume the form of the spatial concentration distribution of the effluent downwind of the source. The usual assumption is that the concentration distribution from a continuous source has a Gaussian distribution relative to the center line in the vertical direction and perpendicular to the wind. For example, effluent from a continuous point source moves downwind, spreading horizontally and vertically such that while the mass in any cross section is constant, the distribution of concentration of pollutant in this cross section along either the horizontal or vertical direction is Gaussian (Wanta, 1968). The standard deviations of the two normal distributions are adjustable and increase with distance or travel time. They are functions of the diffusion coefficients. The standard deviation of the horizontal profile is usually much greater than the vertical. Reflecting planes, such as the ground and an inversion, can be included. Thus, the concentration of an inert material emitted at a rate Q from a continuous point source at a height h is given by (Panofsky, 1969)

$$c(x,y,z) = \frac{Q}{2\pi\bar{u}\sigma_y\sigma_z} \exp\left[\frac{-y^2}{2\sigma_y^2}\right] \exp\left\{-\frac{(z+h)^2}{2\sigma_z^2} - \frac{(z-h)^2}{2\sigma_z^2}\right\} \qquad (7)$$

where $\bar{u}$ is assumed constant. Methods differ in the way in which the empirical standard deviations, σ_y and σ_z, are determined. Equation (7) and the corresponding relation for a line source has seen widespread use in atmospheric diffusion predictions (Gifford, 1968; Turner, 1969).

Review of Urban Atmospheric Models

Several urban atmospheric diffusion models have been proposed. A summary of some of the most important of these models appears in Table I, many of the entries of which have been taken from similar summaries of Wanta (1968) and Lamb (1968). Other models, not appearing in Table I have been proposed by Plato, *et al.,* (1967) and Martin and Tikvart (1968).

At this time the mathematical modeling of atmospheres with inert contaminants is based primarily on the use of exact solutions of (6) or the Gaussian distribution assumption (7). Since no reactions occur, concentrations resulting from different sources are purely additive. Thus, a relation of the form of (7) can be used for each source in the region and the total contributions summed over all the sources.

Prior urban atmospheric diffusion models have had several deficiencies. First, air pollution over many urban areas involves many species which participate in chemical reactions producing secondary contaminants, and, almost exclusively, atmospheric modeling studies have been concerned with passive effluents. Second, the Gaussian distribution assumption (7) has usually been applied to a continuous point or infinite line source perpendicular to the wind direction and has not included the effect of time variations of source strengths and the existence of upper level inversions.

Lamb (1968) has listed the properties that a general atmospheric diffusion model should possess:

1. Inclusion of point, finite line, and area sources with strengths variable in time and space.
2. Time dependence.
3. Variable diffusion coefficients.
4. Winds variable in time and space.
5. Stability and inversion height variable in time and space.
6. Absorption of contaminants by the ground.
7. Allowance for chemical reactions between the various constituents treated.

Undoubtedly, the most difficult aspect to include is item 7, since when chemical reactions take place among the contaminants, because, in general, nonlinear terms are introduced in (6), exact solutions of (6) may no longer be obtained.

The most extensive atmospheric diffusion model to date is the Los Angeles basin model of Lamb (1968). Lamb's model includes the following aspects:

1. Time dependence of concentrations.
2. Point, line, and area sources with strengths variable in space and time.
3. Absorption at the ground.
4. Simple chemical reactions whose rates may be given as arbitrary functions of time.
5. Winds variable in time and horizontally but not vertically.

178

Table I.

Summary of Urban Atmospheric Diffusion Models

	Pooler, 1961	Turner, 1964
Working Equation	$\dfrac{2Q \exp\left[\dfrac{-h^2}{2\sigma_z^2}\right]}{\sqrt{2\pi}(1/8\pi x)u\sigma_z}$	$\dfrac{Q}{\pi \bar{u}\sigma_y\sigma_z} \exp\left[-\dfrac{1}{2}\left(\dfrac{y^2}{\sigma_y^2} + \dfrac{h^2}{\sigma_z^2}\right)\right]$
Types	area modified point source	area modified point source
Size	1 mi. × 1 mi.	1 mi. × 1 mi.
Source Number		272
Time Dependence	1 month	2 hours
Emission height	30 m.	24 m.
Computed Concentrations		
Spatial Resolution	1 mi. × 1 mi.	1 mi. × 1 mi.
Time Resolution	1 month	2 hours
Diffusion Coefficients	$\sigma_z^2 = 2L\bar{u}\alpha - \beta_x\beta$	Empirical σ_y and σ_z as a function of travel time and stability
Wind		
Horizontal Resolution	constant	constant
Vertical Resolution	constant	constant
Time Resolution	1 month	2 hours
Inversion height	none	0 or ∞
Stability Classes		7
Chemical Reactions	Decay of SO_2 in σ_z	Half life of SO_2 = 4 hours
Applied to	Nashville	Nashville
Rollutant	SO_2	SO_2
Remarks	Concentration assumed uniform across horizontal angular sector of 22.5°. One-half of observed pollution at 123 stations within factor of 1.25 of predicted.	58% of calculated 24 hour concentrations at 32 stations within 1 pphm of observed.

	Clarke, 1964	Koogler, et. al., 1967
Working Equation	same as Pooler	same as Turner
Types	area	area
Size		1 mi. × 1 mi.
Source Number	4 sectors	198 area, 45 point
Time Dependence	none	8 hours
Emission height	30 m.	Variable
Computed Concentrations		
Spatial Resolution	2 stations	1 mi. × 1 mi.
Time Resolution	2 hours	1 hour
Diffusion Coefficients	same as Turner	Empirical σ_y and σ_z
Wind		
Horizontal Resolution	constant	constant
Vertical Resolution	constant	$\dfrac{u_1}{u_2} = \left(\dfrac{z_1}{z_2}\right)^{\frac{n}{2} - 2}$
Time Resolution	2 hours	1 hour

Table I. (cont.)

Inversion height	none	arbitrary
Stability Classes	5	5
Chemical Reactions	Half life of SO_2 = 4 hours none for NO_x	Half life of SO_2 = 4 hours
Applied to	Cincinatti	Jacksonville, Florida
Pollutant	SO_2 and NO_x	SO_2
Remarks	14 of 19 calc. 24 hr. conc. of NO_x within 2 pphm of obs.	90% of calculated concentrations within 1 pphm of observed

	Hilst, *et. al.,* 1967	Miller and Holzworth, 1967
Working Equation	same as Turner	$\dfrac{2Q}{\sqrt{2\pi}\,\sigma_z u}$
Types	area	area
Size	$5000' \times 5000'$	variable
Sources Number	5600	
Time Dependence	1 hour	2 hours
Emission height	5 heights possible	surface
Computed Concentrations		
Spatial Resolution	$5000' \times 5000'$	
Time Resolution	1 hour	2 or 4 hours average
Diffusion Coefficients	Empirical σ_y and σ_z	Empirical σ_z
Wind		
Horizontal Resolution	not specified	constant
Vertical Resolution	not specified	constant
Time Resolution	1 hour	2 hours
Inversion height	0 or ∞	variable
Stability Classes	6	2
Chemical Reactions	None	None
Applied to	Connecticut	Los Angeles, Nashville, Washington
Pollutant	SO_2 and CO	SO_2 and NO_x
Remarks	No comparison with observed values	Reasonable agreement with observations in the 3 cities

	Ott, *et. al.,* 1967	Slade, 1967
Working Equation	same as Pooler	see paper (Gaussian with modification)
Types	area	area
Size	variable	variable
Sources Number	24	
Time Dependence	1 hour	none
Emission height	variable	surface
Computed Concentrations		
Spatial Resolution	zonal grid	50 points
Time Resolution	1 year	none
Diffusion Coefficients	same as Pooler	Empirical

Table I. (cont.)

Wind		
Horizontal Resolution	constant	constant
Vertical Resolution	constant	constant
Time Resolution	1 hour	constant
Inversion height	none	800 m.
Stability Classes		1
Chemical Reactions	none	variable half-life
Applied to	Washington and Chicago	Washington to Boston
Pollutant	CO	CO_2
Remarks		No comparison with observed values

	Lamb, 1968
Working Equation	see text
Types	area, finite line and point
Size	variable
Sources Number	198 area, 107 line
Time Dependence	1 hour
Emission height	variable
Computed Concentrations	
Spatial Resolution	variable 1600 × 200 meters
Time Resolution	1 hour
Diffusion Coefficients	Constant K_y and K_z
Wind	
Horizontal Resolution	4800 × 4800 m.
Vertical Resolution	constant
Time Resolution	½ hour.
Inversion height	arbitrary
Stability Classes	1
Chemical Reactions	Arbitrary functions of time
Applied to	Los Angeles
Pollutant	CO
Remarks	see text

6. Constant diffusion coefficients.

7. Constant inversion height.

The model was based on the solution of the diffusion equation,

$$\frac{\partial c}{\partial t} + u(x,y,t)\frac{\partial c}{\partial x} + v(x,y,t)\frac{\partial c}{\partial y} = K\left[\frac{\partial^2 c}{\partial x^2} + \frac{\partial^2 c}{\partial y^2}\right]$$

$$+ K_z\frac{\partial^2 c}{\partial z^2} + f(t)c + S(x,y,z,t) \qquad (8)$$

$$c(x,y,z,0) = \gamma(x,y,z)$$

$$\lim_{x,y \to \pm\infty} c(x,y,z,t) = 0$$

$$K_z \frac{\partial c}{\partial z} = -\beta c \qquad z = H \text{ (ground level)}$$

$$K_z \frac{\partial c}{\partial z} = 0 \qquad z = 0 \text{ (inversion height)}$$

$u(x,y,t)$ and $v(x,y,t)$ are the x and y components of the surface wind, assumed indipendent of z. $\gamma(x,y,z)$ is the initial concentration distribution of pollutant, and β is the first order rate coefficient for absorption of pollutant at the ground. The term $f(t)c$ represents the rate of production of the component by chemical reaction and $S(x,y,z,t)$ is the source function, e.g. a ground-level area source would be represented by $S'(x,y,t)\delta(z-h)$.

Notable approximations in the model are the lack of z dependence and u and v, the lack of inclusion of a vertical wind component, the use of constant values of K and K_z, and a constant inversion height. Nevertheless, this model represents a significant advancement in generality over prior urban diffusion models.

Because of the general functions $f(t)$ and $S(x,y,z,t)$, the solution of (8) is in terms of integral equations, which must be evaluated numerically. In use of the solution of (8), a rectangular array of spatial points at which concentrations are computed is laid out on the region. The x and y components of the surface wind are specified at points on the rectangular grid and a time interval for evaluation of wind components is selected. The sources on the grid emit a puff at each time step in the integration, and the puffs are followed until they are fully dispersed. Finally, the effects of all the source emissions are added to compute concentrations as a function of time and location.

The model was used to compute CO concentrations over Los Angeles on September 23, 1966, at 1200 grid points, resulting from 107 line and 198 area sources (vehicular traffic only). The computed concentrations were compared with concentrations measured at various stations in the basin. It was found that the computed CO concentrations were too high at points of wind trajectory convergence, probably due to lack of inclusion of a vertical wind component or variation of u and v with z. Afternoon computed concentrations were consistently too low, presumably because of sources outside the Los Angeles basin.

In its present form, Lamb's model is the most sophisticated urban diffusion model yet proposed. However, it suffers from one serious drawback. The manner in which chemical reactions have been included is overly restrictive, in that the rates of production and disappearance of reactive species have far more complex dependences on concentration than the f(t)c term in (8). In general, f(t) will not be known, rather only the form of the reaction rates as a function of concentrations will be all that is known. Thus, a more accurate model will require an equation of the form of (8), which is a special case of (6), for each species, each of which is coupled to one or more of the other equations through the reaction rate terms. As noted, this will generally result in nonlinear terms, rendering analytical solutions impossible.

Several other studies have appeared which focus on the problem of modeling the Los Angeles basin, including the effect of reactions.

In an important early paper, Frenkiel (1956) incorporated a Gaussian diffusion model and time variations of the wind in an urban atmospheric model. The city was divided into a number of subareas, and each pollutant subcloud, initially the size of the subarea source, grows by diffusion at some rate. The average concentration in each subcloud decreases as the cloud moves away from its origin, but the contribution from each cloud is added to give pollutant concentrations as a function of time and position over the whole city. He estimated the relative contributions of automobiles, industry and other sources to Los Angeles atmospheric O_3 concentrations as a function of time of the day at selected locations. His calculations were based on certain simple assumptions on the kinetics of O_3 formation. It was found that due to the chemical reactions, the contributions of the various sources to O_3 formation in the Los Angeles basin are not simply additive. One point growing out of Frenkiel's study was the need for a good kinetic model of the atmospheric reactions.

Neiburger (1959) identified the key meteorological problem associated with Los Angeles smog, namely light winds and an inversion. He showed that air trajectories in the Los Angeles basin consist of alternate large displacements landward during the day followed by small displacements seaward at night. He traced the trajectory of an air parcel reaching Pasadena at noon and computed the amount of hydrocarbon added to the parcel as it moved along its trajectory. The calculation clearly established that automobile traffic is the major source of hydrocarbons in the Los Angeles atmosphere.

A detailed study of meteorology and Los Angeles smog has been performed by Shuck, et al., (1966). The concentration of maximum daily oxidant was found to be a function of day of the week and related to automobile traffic patterns. Weekend oxidant and temperature effects suggested that the magnitude of local pollution is more important than horizontal transport in determining the weekly pattern of smog symptoms at any given station. The absolute magnitude of pollutant levels are determined by horizontal transport and inversion height and are proportional to each other over large areas of the basin. The severity of smog was found to be related to early morning pollution levels and the ratio of hydrocarbons to oxides of nitrogen. Data analyzed indicated that control of the intensity of certain smog symptoms is directly proportional to the control of hydrocarbons, and that control of oxides of nitrogen, to be equally effective, must be greater than 50 percent. For an atmospheric model, the authors assumed that the Los Angeles basin acts as a large irradiation chamber with the major sources of pollutants evenly scattered throughout the chamber.

Apparently the first systematic investigation of the control of air pollution in the Los Angeles basin has been made by Ulbrich (1967, 1968). The control strategy utilized was termed "adapredictive" by Ulbrich and represents a combination of adaptive and predictive control. The outputs from a real-time process model are continually compared to those from the real system (the atmosphere) for use in adapting the model to more closely fit the true system. Once this adaptive scheme has defined an adequate process model, then the validated model is used in fast time on a computer to predict what the true system output will be as a result of some specific control strategy. The Los Angeles basin was divided into seven sections, each of which would have a measurement station the readings from which would be sent to the central computer system. It was assumed that fast homogeneous mixing of pollutants takes place throughout the zone of generation and that the pollutants are transported horizontally by the wind without significant mixing above the inversion. Although the overall computer control scheme outlined by Ulbrich represents the ultimate framework in which an atmospheric model will be placed, the particular atmospheric model utilized in that study was highly over-simplified, particularly with respect to the chemistry of smog formation.

Mathematical Models of Chemically Reacting Air Pollution

The question we want to consider in this section is the construction of the atmospheric model for air pollution which may be chemically reacting which will ultimately be an integral part of an air monitoring-computer simulation system that provides predictions on the state of the atmosphere depending on the levels of the uncontrollable and controllable inputs.

In theory, the complete concentration-time history of an air quality control region can be obtained by solution of (6) for each of the N species, subject to the appropriate boundary conditions. However, the simultaneous solution of several of these partial differential equations on a spatial domain representing an urban area is a formidable task for even the largest of modern computers. In addition, lack of detailed knowledge of reaction rate constants for atmospheric reactions, (not to mention the reaction mechanism itself), wind patterns, source strengths, atmospheric turbulence characteristics in the presence of buildings, values of K_x, K_y and K_z, etc., makes an approach based on the numerical solution of equations of the type (6) highly unnecessary and impractical. Thus, we will concentrate on more simplified models which are computationally feasible. As more knowledge is gained about atmospheric reactions and dispersion, models based on the numerical solution of (6) may become more feasible.

We will consider two different types of mathematical models for chemically reacting air pollution. The first type of model will be termed *extended similarity models*, in which models like those in Table 1 and that of Frenkiel are used, coupled with information on the rate of chemical reactions occurring. Thus, in this type of model, a source is identified and its effluent is followed as it is dispersed by the wind. Some sort of combination of the effluent clouds from all sources provides average concentrations in the urban area.

The second class of atmospheric models involves division of the air quality control region into a number of separate regions, each of which behaves as a well-mixed cell. This model, which we will call the *well-mixed cell model,* is based on air volumes of fixed size, each of which has a uniform pollutant concentration. This concept has already been employed by Ulbrich (1967) in his study of the Los Angeles basin.

Extended Similarity Models

In the case of a passive effluent the specification of the form of the concentration distribution, as in (7), rather than solution of (6), has proved to be a highly successful approach. A logical extension to the reactive species case would thus be to assume some form for the concentration distributions that includes the effect of chemical reactions.

The hypothesis that turbulent motions of particles in steady, self-preserving, free shear flow possess similarity in the Lagrangian sense was proposed by Batchelor (1957, 1964). Later, Lagrangian similarity arguments, i.e. that motion of a marked particle in turbulent shear flow may be similar at stations downstream from the point of release, were applied to diffusion data obtained in the laboratory and in the surface layer of the atmosphere by Cermak (1962). Chatwin (1968) considered the dispersion of a cloud of passive material released from an instantaneous source in the constant stress region. He assumed that the vertical turbulent transfer could be described by an eddy diffusivity and that the Lagrangian similarity hypothesis is applicable.

Let us consider the case of an instantaneous point source at height h above ground level, from which a particle of inert contaminant is released into the turbulent atmospheric surface layer. If the mean position of the particle at any time t is $\bar{x}(t)$, $\bar{y}(t) = 0$, $\bar{z}(t)$, according to the Lagrangian similarity hypothesis the average concentration at any point in a cloud of particles is given by

$$c_i = \frac{Q_i}{\bar{z}^3} \, \psi \left[\frac{x-\bar{x}}{\bar{z}} , \, \frac{y}{\bar{z}} , \, \frac{z-\bar{z}}{\bar{z}} \right]$$

$$= \frac{Q_i}{\bar{z}^3} \, \psi \left[n_x, n_y, n_z \right] \tag{9}$$

where Q_i is the mass of species i emitted at $t = 0$ and ψ is a general probability density function. The relation is expected to hold for $t > h/u_*$ where u_* is the friction velocity.

Let us assume that the same form can be used in the case of a reacting species but with Q_i a function of time. This assumption appears reasonable if the characteristic time for chemical reaction is long compared to the characteristic time for the mixing processes. In fact, it can be shown that if the fractional rate of change of material due to chemical reaction is small compared to the fractional rate of expansion of the cloud,

$$\left| \frac{\dfrac{1}{Q_i} \dfrac{dQ_i}{dt}}{\dfrac{1}{\bar{z}} \dfrac{d\bar{z}}{dt}} \right| \ll 1$$

the similarity solution will apply (Friedlander and Seinfeld, 1969).

For an instantaneous point source let Q_{i_0} be the mass of species i emitted, and $Q_i(t)$ be the mass of species i in the could at time t. The total amount of species i changes only by chemical reaction,

$$\frac{dQ_i}{dt} = \int_0^\infty \int_{-\infty}^\infty \int_{-\infty}^\infty R_i \, dx \, dy \, dz \tag{10}$$

In general, $R_i = R_i(c_1, \ldots, c_N)$, and for simple reactions,

$$R_i = \sum_{j=1}^M \sigma_{ij} k_j \prod_{\ell=1}^N c_\ell^{\alpha_{\ell j}} \tag{11}$$

where M is the number of reactions and σ_{ij} are the stoichiometric coefficients. Combining (9), (10) and (11), and changing the integration variables to n_x, n_y, and n_z, we obtain

$$\frac{1}{\bar{z}^3} \frac{dQ_i}{dt} = \sum_{j=1}^M \sigma_{ij} k_j \prod_{\ell=1}^N \left[\frac{Q_\ell}{\bar{z}^3} \right]^{\alpha_{\ell j}} \int_{-1}^\infty \int_{-\infty}^\infty \int_{-\infty}^\infty$$

$$\psi(n_x, n_y, n_{\bar{z}})^{\sum_{\ell=1}^N \alpha_{\ell j}} \quad dn_x dn_y dn_z \tag{12}$$

186

If we define the dimensionless parameters,

$$A_j = \int\limits_{-1}^{\infty} \int\limits_{-\infty}^{\infty} \int\limits_{-\infty}^{\infty} \psi(n_x, n_y, n_z)^{\sum\limits_{\ell=1}^{N} \alpha_{\ell j}} \, dn_x dn_y dn_z \qquad (13)$$

then (12) becomes

$$\frac{1}{\overline{z}^3} \frac{dQ_i}{dt} = \sum_{j=1}^{M} \sigma_{ij} k_j A_j \prod_{\ell=1}^{N} \left[\frac{Q_\ell}{\overline{z}^3} \right] \qquad (14)$$

Diffusion and chemical reaction effects have become partially uncoupled. The term $Q_i/\overline{z}^3$ is equivalent to the concentration of component i in the cloud and $\overline{z}^3$ to the cloud volume. Diffusional aspects appear in the rate at which the cloud volume expands with time, $\overline{z}^3(t)$, which depends on the meteorological condition. The cloud as pictured in Figure 2 can be considered as a batch reactor with variable volume, the reaction rates governed by kinetics and the rate of volume change governed by meteorology. The concentration at ground level is is determined from

$$c_i = \frac{Q_i}{\overline{z}^3} \psi(0,0,-1) \qquad (15)$$

Since $\overline{z}$, the mean position of the cloud above ground, increases as the cloud expands, the ground-level concentration of a non-reacting species, such as CO, decreases continuously as the cloud expands. For a species generated by a chemical reaction, such as NO_2 or O_3, the ground-level concentration could increase with time for some portion of the cloud trajectory than decrease. For an irreversible bimolecular reaction, mixing not only reduces the reaction rate (dependent on the concentrations of the two components) but also the ultimate amount of product formed. Thus, atmospheric conditions in which rapid mixing occurs tend to reduce both the rate of production and the ultimate quantities of secondary contaminants.

As $\overline{z} \to \infty$ the maximum amount of smog products formed is always less than the maximum amount of these products which would be produced in a constant volume batch reactor. The reduction results from the quenching effect of the rapid expansion of the cloud.

The Lagrangian similarity hypothesis can also be extended to diffusion and slow chemical reaction in turbulent shear flows from continuous sources. For a continuous point source, let W_{i_0} be the rate of emission of species i at

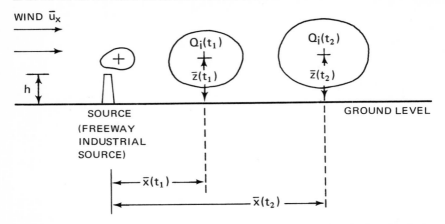

WIND $\bar{u}_x$

$Q_i(t_1)$

$\bar{z}(t_1)$

$Q_i(t_2)$

$\bar{z}(t_2)$

h

SOURCE
(FREEWAY
INDUSTRIAL
SOURCE)

GROUND LEVEL

$\bar{x}(t_1)$

$\bar{x}(t_2)$

An instantaneous point source of reactants forms a chemically-reacting cloud which expands because of convection and diffusion. It can be conxidered as a well-mixed reactor of variable volume.

Q_i = Total mass of component i in cloud.

Figure 2.

the source (mass/time) and $W_i(x)$ be the rate that species i crosses a plane at constant x (mass/time). At any plane x,

$$W_i(x) = \int_0^\infty \int_\infty^\infty \bar{u}(z) \; c_i(x + \Delta x, y, z)\,dy\,dz \qquad (16)$$

and at a plane $x + \Delta x$,

$$W_i(x + \Delta x) = \int_0^\infty \int_\infty^\infty \bar{u}(z) \; c_i(x + \Delta x, y, z)\,dy\,dz \qquad (17)$$

Subtracting (16) from (17) and taking the limit as $\Delta x \to 0$,

$$\frac{dW_i}{dx} = \int_0^\infty \int_\infty^\infty \bar{u}(z) \; \frac{\partial c_i(x,y,z)}{\partial x} \; dy\,dz \qquad (18)$$

The appropriate form of (6) for the continuous point source is

188

$$\bar{u}(z)\frac{\partial c_i}{\partial x} = \frac{\partial}{\partial y}\left[K_y \frac{\partial c_i}{\partial y}\right] + \frac{\partial}{\partial z}\left[K_z \frac{\partial c_i}{\partial z}\right] + R_i \qquad (19)$$

$$K_y \frac{\partial c_i}{\partial y} = 0 \qquad y \to \pm \infty$$

$$K_z \frac{\partial c_i}{\partial z} = 0 \qquad z \to 0, \infty$$

Combining (18) and (19) and integrating, we obtain

$$\frac{dW_i}{dx} = \int_0^\infty \int_\infty^\infty R_i \, dy \, dz \qquad (20)$$

Thus, the change in the total amount of species i in the downwind direction is due only to chemical reaction.

The Lagrangian similarity hypothesis for the continuous point source is

$$c_i(x,y,z) = W_i(x) \int_0^\infty \frac{\psi}{\bar{z}^3}(n_x, n_y, n_z) \, dt \qquad (21)$$

If the integral in (21) is denoted $\Lambda(x,y,z)$, then (21) can be combined with (20) and (11) to yield

$$\frac{dW_i}{dx} = \sum_{j=1}^M \sigma_{ij}k_j B_j(x) \prod_{\ell=1}^N W_i(x)^{\alpha\ell j} \qquad (22)$$

where

$$B_j(x) = \int_0^\infty \int_\infty^\infty \Lambda(x,y,z)^{\sum_{\ell=1}^N \alpha\ell j} \, dy \, dz \qquad (23)$$

For a continuous line source, let V_{i_0} be the rate of emission of species i per unit length (mass/time-length) and $V_i(x)$ be the rate that species i cross a plane at constant x (mass/time-length). At any plane x,

189

$$V_i(x) = \int_0^\infty \bar{u}(z) \, c_i(x,z) dz \qquad (24)$$

The change in the total mass of species i with distance x is governed by

$$\frac{dV_i}{dx} = \int_0^\infty R_i \, dz \qquad (25)$$

The Lagrangian similarity hypothesis for a continuous line source is

$$c_i(x,z) = V_i(x) \int_0^\infty \int_\infty^\infty \frac{\psi \, (n_x, n_y, n_z)}{\bar{z}^3} \, dy \, dt \qquad (26)$$

Denoting the integral in (26) by $\Omega(x,z)$, (26) can be combined with (25) and (11) to give

$$\frac{dV_i}{dx} = \sum_{j=1}^{M} \sigma_{ij} k_j C_j(x) \prod_{\ell=1}^{N} V_i(x)^{\alpha_{\ell j}} \qquad (27)$$

where

$$C_j(x) = \int_0^\infty \Omega(x,z)^{\sum_{\ell=1}^{N} \alpha_{\ell j}} \, dz \qquad (28)$$

For continuous sources $B_j(x)$ and $C_j(x)$ contain the meteorological information, namely the dependence of $\bar{z}$ on x. Thus, solutions of the ordinary differential equations (22) and (27) can be used in the same way as predictions from Gaussian plume equations are used in the inert case.

One drawback of the Lagrangian similarity approach is that the exact form of ψ has yet to be determined. An obvious approximate choice for ψ would be the Gaussian probability density.

Another problem yet to be adequately answered is: what happens when the chemically reacting effluents from two sources intersect? Because the chemical reaction rates are nonlinear functions of the concentrations, we know that the reaction rates in the overlapping region are not simply the sum of the reaction rates in the individual plumes. This question is currently being studied.

Well-Mixed Cell Model

This approach has several advantages. First, detailed wind trajactories need not be known, rather only an estimate of the intercell flows. Second, all the sources in each cell are considered as contributing uniformly throughout the volume of the cell so that lack of complete knowledge of individual source strengths may not be a serious drawback. Third, since the dynamic behavior of the pollutant in each cell is governed by ordinary differential equations, a high-speed computer simulation will be capable of treating many cells efficiently and rapidly. The well-mixed cell model will be particularly applicable when rapid vertical mixing takes place under an inversion layer. On the other hand, when an inversion is very high or does not exist, this approach may not give a valid description of the atmosphere.

Let us consider an air quality control region which has been divided into L cells, each of which may be considered to be a well-mixed reactor. The volumes of the cells, which need not be equal, will be denoted $v_1, \ldots, v_L$. The concentration of species i in cell j will be x_{ij}. In each cell there will be a time-varying source of each pollutant, the rate of emission of species i into cell j being S_{ij}. Also, there exists the possibility that pollutants can be removed in each cell, the rate of removal of species i from j being D_{ij}. Finally, the volumetric rate of airflow from cell j to cell k will be q_{jk}.

The simplest case of the well-mixed cell model is to describe the entire air quality control region by a single cell (L = 1). In this case, if $\bar{u}$ is the wind speed normal to one of the sides, H is the inversion height, S the rate of introduction of an inert contaminant per area, and d the length of a side for a square area, then Smith (1961) has shown that the equilibrium concentration of the pollutant is

$$x_e = \frac{Sd}{uH} \tag{29}$$

which is 90 percent reached in $t = 2.3 \, d/\bar{u}$. Such an approach is obviously too oversimplified for use in a mathematical model.

Let us now consider a dynamic material balance for species i in cell j. The contributions to the change in concentration of species i in cell j can be listed:

$$\sum_{j=1}^{L} q_{jk} x_{ij} = \text{inlet flow of i from all other cells, mass/time} \quad (q_{kk} = 0)$$

$$x_{ik} \sum_{j=1}^{L} q_{kj} = \text{outlet flow of i from cell k to all other cells, mass/time}$$

191

S_{ik} = input of i from sources in cell k, mass/time

D_{ik} = removal of i by sinks in cell k, mass/time

R_{ik} = rate of production of i by chemical reaction in cell k, mass/time

Thus, a dynamic balance for species i in cell k is

$$V_k \frac{dx_{ik}}{dt} = \sum_{j=1}^{L} q_{jk}x_{ij} - x_{ik} \sum_{j=1}^{L} q_{kj} + S_{ik} - D_{ik} + R_{ik} \qquad (30)$$

The air quality region model consists of equations of the form (30) for each of the N components in each of the L cells, i.e. LN ordinary differential equations. For example, for a model of the Los Angeles basin the species of interest might be NO, NO_2, and hydrocarbons (N = 3). A division into 10 cells would necessitate the integration of 30 ordinary differential equations. To begin using the model, an initial time and set of initial cell concentrations would be selected to provide initial conditions for (30).

One drawback of the well-mixed cell approach is that if the cells are fairly large, local meteorological effects will not be included in the model. Thus, in determining the number of cells to represent an air quality region a trade off must be made between a large number of cells for accurate atmospheric representation and the increased cost of computing as the number of cells increases.

Control of Air Pollution

The ultimate objective of a mathematical model of an air quality control region is the control of air pollution. The information flow diagram in Figure 3 depicts the control situation for photochemical smog. The uncontrollable inputs are essentially the weather and topographical conditions, which, for the most part, cannot be changed. The control variables, the quantities which can be changed to produce a desired result, relate to motor vehicle emissions, quantities, constituents, and timing (e.g. rush hours). In addition, approaches such as rapid transit could be included as control variables. Engine and fuel modifications, catalytic and thermal afterburners for the exhaust manifold are means of effecting emission controls. The system is the atmosphere, the output of which is smog. We have available two basic strategies for the control of smog. Feedforward control consists of taking action a priori given knowledge of the uncontrollable inputs and how bad the smog will be as a result. The setting of automotive emission standards by a legislative body is in an

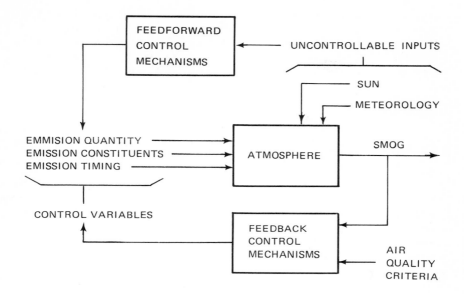

CONFIGURATIONS FOR CONTROL OF SMOG

Figure 3.

extended sense feedforward control. Obviously, for effective feedforward control one needs a reliable model of the system. Feedback control consists of taking action based on the current system output to adjust the control variables. For example, we would take action during the day based on measurements characterizing the severity of smog. The system of stages of smog alerts now existing in Los Angeles is an example of feedback control. Although a more inexact model of the system can be tolerated in feedback as opposed to feedforward control, the large time delays and political implications inherent in feedback control of smog limit its usefulness for the eventual control of smog.

The application of concepts from optimization and control to problems of the air environment is a natural extension of the modeling effort. In particular, the following problems appear to be of substantial importance:

1. The determination of the optimum initial reactant ratios in the photochemical smog system such that some criterion, perhaps the maximum concentration of ozone reached, is minimized.
2. The optimal allocation of primary pollutant sources in an air quality control region such that a criterion based on air quality is maximized.

Summary

The considerations important in the development of mathematical models for air quality control regions have been outlined. Initially, the role of meteorology and the nature of any atmospheric chemical reactions should be discerned. Then these elements are combined in the model, two possible types of which have been detailed. The first, based on the Lagrangian similarity hypothesis, has an assumed form for the distribution of concentration of a reacting species downwind of a source, much in the same way as Gaussian plume equations are used for inert contaminants. The second type of atmospheric model is based on representing the region by a fixed network of well-stirred cells.

A topic which we have not considered is the role of aerosols in atmospheric chemical reactions and air pollution. Ultimately, the effect of aerosols should be incorporated into a mathematical model. Then, perhaps, the visibility reduction in air pollution can be quantitatively described.

Obviously, what has been presented here is only a beginning. An extensive program of model revision and experimental validation must accompany the development of any reliable urban atmospheric model. Thus, the acquisition of accurate urban pollutant concentration and meteorological data is an important link in the modeling effort.

References

Batchelor, G.K., J. Fluid Mech., *3*, 67 (1957).

Batchelor, G.K., Archiwum Mechaniki Stosowanej, *16*, 3, 661 (1964).

Bird, R.B., W.E. Stewart, and E.N. Lightfoot, "Transport Phenomena," Wiley, New York, 1960.

Briggs, G.A., in "Meteorology and Atomic Energy," D.H. Slade, Ed., U.S. Atomic Energy Commission, TID-24190, July 1968.

Cermak, J.E., J. Fluid Mech., *15*, 49 (1963).

Chatwin, P.C., Quart. J. Royal Meteorol. Soc., *94*, 401, 350 (1968).

Clarke, J.F., J. Air Poll. Control Assoc., *14*,9, 347 (1964).

Frenkiel, F.N., Smithsonian Inst. Ann. Rept., 296 (1956).

Friedlander, S.K., and J.H. Seinfeld, "A Dynamic Model of Photochemical Smog," Environ. Sci. Tech., in press.

Gifford, F.A., in "Meteorology and Atomic Energy," D.H. Slade, Ed., U.S. Atomic Energy Commission, TID-24190, July 1968.

Koogler, J.B., R.S. Sholtes, A.L. Davis, and C.I. Harding, J. Air Poll. Control Assoc., *17*, 4, 211 (1967).

Lamb, R.G., "An Air Pollution Model of Los Angeles," M.S. Thesis, University of California, Los Angeles, 1968.

Leighton, P.A., "Photochemistry of Air Pollution," Academic Press, New York, 1961.

Martin, D.O., and J.A. Tikvart, "A General Atmospheric Diffusion Model for Estimating the Effects of One or More Sources on Air Quality," National Air Pollution Control Administration, Cincinatti, 1968.

Miller, M.E., and G.C. Holzworth, J. Air Poll. Control Assoc., 17, 1, 46 (1967).

Neiburger, M., in "The Rossby Memorial Volume," Rockerfeller Institute Press, New York, 1959.

Panofsky, H.A. Amer. Scientist, 57, 2, 269 (1969).

Pasquill, F., "Atmospheric Diffusion," D. Van Nostrand, Dondon, 1962.

Plato, P.A., D.F. Menker, and M. Dauer, Health Physics, 13, 1105 (1967).

Pooler, F., Jr., Int. J. Air Water Pollution, 4, 199 (1961).

Saltzman, B.E., Ind. Eng. Chem., 50, 4, 677 (1958).

Schuck, E.A., J.N. Pitts, and K.S. Wan, Int. J. Air Water Pollution, 10, 689 (1966).

Slade, D.H., Science, 157, 1304 (1967).

Smith, M.E., Intern. Symp. Chem. Reactions in the Lower and Upper Atmosphere, Stanford Research Institute, San Francisco, 1961.

Stephens, E.R., "Chemistry of Atmospheric Oxidants," presented at 61st Annual Meeting of the Air Pollution Control Association, St. Paul, June 1968.

Turner, D.B., J. Appl. Meteorol., 3, 83 (1964).

Turner, D.B., "Workbook of Atmospheric Dispersion Estimates," U.S. Dept of Health, Education and Welfare, National Air Pollution Control Administration, 1969.

Ulbrich, E.A., "Adapredictive Air Pollution Control for the Los Angeles Basin," presented at Assoc. for Computing Machinery Annual Symposium, Applications of Computers to the Problems of Urban Society, November 10, 1967.

Ulbrich, E.A., "A Hybrid Computer Simulation Investigating the Cost Effectiveness of Air Pollution Control Over a 10-Year Period," presented at 61st Annual Meeting of the Air Pollution Control Association, St. Paul, June 1968.

Wanta, R.C., in "Air Pollution," Vol. I, 2nd ed., A.C. Stern, Ed., Academic Press, New York, 1968, Chapter 7.

Wayne, L.G., "The Chemistry of Urban Atmospheres," Technical Progress Report III, Los Angeles County Air Pollution Control District, December 1962.

Wayne, L.G., and T.E. Earnest, "Photochemical Smog, Simulated by Computer," presented at 62nd Annual Meeting of the Air Pollution Control Association, New York, June 1969.

Discussion of Paper, "Mathematical Models
of Air Quality Control Regions," by John H. Seinfeld

By Lowell G. Wayne
Air Pollution Control Institute
School of Public Administration
University of Southern California
Los Angeles, California
and Principal Consulting Investigator
System Development Corporation
Santa Monica, California

Reviewing the types of air quality models Dr. Seinfeld has encountered, I see that in his view they reduce to two principal varieties: one in which some specific form is assumed for the distribution of concentration of a species downwind of a source, and another in which a region is represented as a fixed network of well-mixed cells.

I venture to suggest that neither of these approaches will soon generate mathematical models that will be of much help in understanding the behavior of photochemical smog. In contrast to dynamic simulation models, the mathematical models suffer from an over-emphasis on the development of patterns, preferably patterns which represent approximate analytical solutions of diffusion equations. Given the present status of mathematical knowledge, this endeavor requires the denial of important aspects of chemical reality, as Dr. Seinfeld has shown. The usual procedure has been to develop equations for describing the behavior of pollutants which do not react in the atmosphere, and then apply minor adjustments to represent some sort of "allowance" for chemical reaction.

Thus, in the "adapredictive control strategy" advocated by Ulbrich, the outputs from a model are continually compared to data from the atmosphere for use in adapting the model to fit "the true system." But, surely, no amount of tinkering will make a model fit the atmosphere unless the real chemical behavior of contaminants in the atmosphere is allowed to enter the model in a basic way in the first place.

If a model is to be adjusted to fit the real findings of a monitoring network, there will have to be choices as to what variables are to be included as input and which will emerge as output, and how much aggregation is permissible. Any "adapredictive" model shares the drawbacks of regression models to some extent. These include:

(1) problems in rationalizing the manipulable mathematics;
(2) problems in accounting for observed variance;
(3) problems in choosing appropriate variables, both dependent and independent.

Thus the model arrived at by Schuck, Pitts and Wan, on the basis of statistical averaging, led to the conclusion that "the absolute magnitude of pollutant levels — are proportional to each other over large areas of the basin,"

Yet other statistical approaches point to different conclusions; for example, Figure 1, taken from Los Angeles data, indicates that, for a nine-station network in the basin, maximum ozone levels are highest where maximum NO_x levels are lowest, and vice-versa.

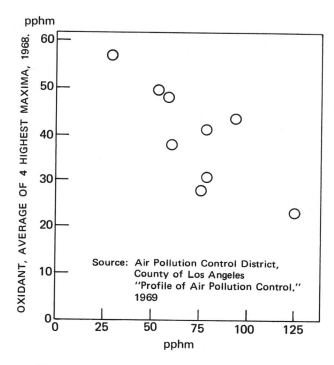

Figure 1. NO_x: average of 4 highest maxima at various monitoring stations in LA Basin, 1968.

The ideal model must be based on physical principles. Studies of contaminant distribution and statistical behavior show (1) important variations in average contaminant concentrations from place to place in the Los Angeles Basin; (2) important variations in the average daily peaks; (3) important variations in the ratios of averages, and of average peaks; etc. These are consistent with the supposition that the most prevalent smog pattern in the Basin is one in which contaminated air is irradiated by the sun as it is transported inland by the sea breeze, and the secondary contaminants develop after an appropriate period of irradiation in whatever localities they have by then reached. There appear (from LACAPCD figures) to be three readily distinguishable types of air quality, associated mainly with distance of travel of air from the shore: a band adjacent to the shore having high average values of primary contaminants, but low secondary contaminants; a band next to the mountains having relatively low levels of primary contaminants and high secondary contaminants; and an intervening central band, intermediate in most respects.

This being the case, it is clear that useful models cannot be built on average behavior, but must respond to really possible behavior; that is to typical trajectories rather than average ones.

Similarly, in specifying the effect of concentration changes on rates of photochemical changes in the atmosphere, it is necessary to remember that the behavior of a mixture may be very different from the "average" behavior of unmixed reactants. As a case in point: much has been made of the fact that certain types of graphs drawn to present information about photochemical experiments sometimes show maxima. Thus, for a series of experiments on irradiation of propylene and nitric oxide diluted in air, the ozone concentration after a particular period for a given hydrocarbon concentration may be at a maximum for a certain initial concentration of nitric oxide, which is found from the experiments. Between this pair of initial concentrations, then, there is a ratio; in some circles, this ratio has been invested with an almost occult significance.

To me, this seems self-defeating. It is clear from the experiments that the value found for this ratio depends on many factors; among them, the initial concentration of the hydrocarbon, the particular hydrocarbon chosen, the type of hydrocarbon chosen, the intensity of irradiation provided, the period of irradiation selected and — most important — the criterion selected

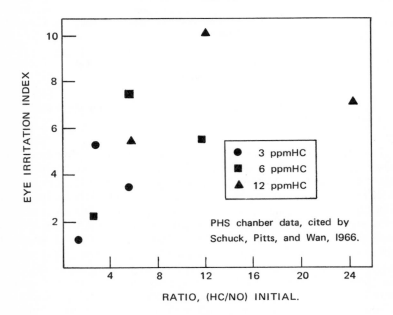

Figure 2.

for the maximization. If a new set of tests is done with any of these factors changed, then the position of the maximum changes (or, indeed, the maximum may vanish entirely) and the corresponding HC/NO ratio changes.

But if a different instrument is used to measure the hydrocarbons, the ratio changes. If the relative concentrations of different hydrocarbons change (as they do, every day), the ratio stands still while the photochemistry changes. In either case, the results from the model can be expected to diverge from the appropriate values, and the only remedy is to retain and utilize all available detail as to the composition of the primary contaminants in smog.

In this connection, Schuck, Pitts and Wan report, "In Figure 16 both atmospheric and laboratory data predict the existence of an optimum hydrocarbon to oxides of nitrogen ratio in terms of eye irritation or percent of days resulting in smog." But, as can be seen from Figure 2, when the value of the criterion is plotted against the HC/NO ratio (this was not done by the authors), the maximum appears at a different ratio for each hydrocarbon level. Interpretation, as the authors admit, is difficult.

I therefore differ with Dr. Seinfeld on the proposition that it is important to determine "optimum initial reactant ratios in the photochemical smog system." I think it unlikely that any "optimum" can be found and, therefore, I suggest that we explore, instead, the effects of varying reactant concentrations on the photochemical system.

I am pleased to report that, at System Development Corporation, the construction of an atmospheric simulation model embodying these concepts is already in progress. This work is supported by the Bureau of Criteria and Standards of the National Air Pollution Control Administration, and publication of the details is expected in 1970.

Discussion of Dr. Seinfeld's Paper on
"Mathematical Models of Air Quality Control Regions"

By M. Neiburger

Dr. Seinfeld presents very effectively the purposes and usefulness of mathematical modeling of an urban area. Briefly, an accurate model would enable prediction of the concentration at all points in and around the area resulting from a given distribution of pollution sources under given meterological conditions. It would enable, for instance, determining whether under adverse meteorological conditions the air quality standards desired could be met with existing sources of emission, and if not, how much reduction of emission rates would be required to achieve them.

It is gratifying to me that Dr. Seinfeld speaks highly of the model designed by my student, Robert Lamb. Lamb has undertaken the task of dealing with the complete problem in a succession of phases. In the phase which he reported in his Master's thesis he dealt essentially with the diffusion problem, dealing with the meteorological aspects in almost complete generality, but taking into account chemical reactions in a highly oversimplified fashion, as Dr. Seinfeld points out. He will subsequently try to take the chemical reactions into account more realistically, as well as to remove the remaining constraints on the meteorological conditions considered.

Apparently Dr. Seinfeld has not appreciated the reasons Lamb used the approach he did. All previous attempts at urban diffusion models which try to deal with the details of the diffusion process have used equation (7) quoted by Seinfeld. This equation expresses the form of the diffusion from a point source under steady unvarying meteorological conditions: wind constant with time and space, and temperature structure similarly constant. It does not allow, for instance, for the kind of diurnal oscillation of wind we have in Los Angeles, with sea breezes in the afternoon, and land winds at night and early morning. To allow for such changes with time, and also variation from the coast inland, of wind direction and speed, one must use the more fundamental equation (6).

The "Extended Similarity Model," being based on an analog to equation (7), has this difficulty, and, it seems to me, the difficulty is compounded since the chemical reaction rates continuously must be computed from incorrect concentrations, since the diffusion is not dealt with correctly.

The deficiencies of the similarity approach are well stated by Dr. Seinfeld. "One drawback of the Lagrangian similarity approach is that the exact form of ψ has yet to be determined . . .

"Another problem yet to be adequately answered is: what happens when the chemically reacting effluents from two sources intersect?"

Indeed, the form of ψ, the probability density function, is the objective of the solution of equation (6), and thus is the crux of the diffusion problem for non-reacting species. For steady conditions the assumption of a Gaussian

200

distribution may be acceptable. For varying wind and stability, and especially for the light winds and strong inversions of adverse meteorological situations, this assumption does not hold.

The condition that the similarity solution apply is stated as "the fractional rate change of material due to chemical reaction is small compared to the fractional rate of expansion of the cloud." As an example, let us consider the reaction

$$NO_2 + h\nu \rightarrow NO + O$$

which is slower than others in photochemical smog. The rate at which this reaction takes place is given by

$$\frac{d\,[NO_2]}{dt} = k\,[NO_2]$$

where k depends on the radiation intensity. For noon solar radiation intensity Leighton gives the value $k \cong 22\ hr^{-1}$.

The expansion of a cloud by diffusion has been shown theoretically and empirically to have the form

$$\overline{Z} = a(\overline{u}\,t)^b$$

where $\overline{u}$ is the mean wind speed, and b has values in the range 0.5 to 1.5, depending on the value of $x = \overline{u}\,t$ and the meteorological conditions. From this expression we get

$$\frac{1}{\overline{Z}}\frac{d\overline{Z}}{dt} = b\,t^{-1}\ ,$$

and

$$\left| \frac{\dfrac{1}{[NO_2]}\dfrac{d\,[NO_2]}{dt}}{\dfrac{1}{\overline{Z}}\dfrac{d\overline{Z}}{dt}} \right| = \frac{k\,t}{b}\ .$$

For this ratio to be less than 0.1, it is necessary that t be less than 0.1 b/k, which in this instance requires that t be less than 25 seconds. Thus even disregarding other concerns about the validity of the procedure, it can be applied only for a fraction of a minute.

In addition, however, the diffusion does not allow one to regard the initially separate volumes of cloud as distinct. With the passage of time they are admixed, and the concentration will be the result of this admixture.

201

From this standpoint Seinfeld is not correct in saying "Diffusion and chemical reaction effects have been partially uncoupled." The reaction rates depend on the concentrations, and the concentrations are determined in large part by the process of diffusion (i.e., the function ψ in Seinfeld's equations). The two types of processes will have to be dealt with simultaneously in order to achieve a realistic evaluation of concentrations.

Comments on *Mathematical Models of Air Quality Control Regions* by John H. Seinfeld

By E. R. Stephens

Statewide Air Pollution Research Center
University of California
Riverside, California 92502

Dr. Seinfeld's paper divides itself rather neatly into two parts, one part dealing with photochemistry of smog and the second and larger part with atmospheric diffusion. This reviewer's principal interest and competence lies in the former so the latter will be left to other reviewers. The photochemical system is a fascinating one, full of paradoxical situations which ambush the unwary investigator. To cite a few examples: the principal hallmark and index of photochemical smog is ozone (O_3), present at a few tenths of a ppm concentration. Ozone reacts so rapidly with nitric oxide (NO) that the two cannot coexist in the same atmosphere at measurable concentrations. But NO is known to be a principal contaminant in auto exhaust and other combustion gases. Superficially it would appear that the hundreds of tons of nitric oxide emitted to an urban atmosphere would quickly scavenge any ozone present! Happily this paradox was resolved about ten years ago, almost before it was fully recognized as a paradox. However the understanding which emerged showed that the concentration of ozone (perhaps the most serious health menace in smog) is *inversely* (rather than directly) related to the concentration of NO (one of the principal causes). This is the steady state equation $[O_3] = k_1 [NO_2]/k_3 [NO]$ first clearly stated by Prof. P. A. Leighton and used by many authors since including Seinfeld. This inverse relationship has raised many questions such as the practical wisdom of controlling NO emissions. It also adds to the challenge of mathematical simulation of the atmosphere.

Simulation of smog formation by irradiation of dilute auto exhaust revealed two additional related paradoxes. Under irradiation the NO initially present is rapidly converted to nitrogen dioxide (NO_2) in spite of the fact that the reverse occurs when NO_2 is present alone. With air and hydrocarbon present nearly all of the NO appears as NO_2. When the NO is gone ozone appears in the mixture and still another surprising thing happens: the nitrogen dioxide just formed immediately begins to disappear. It is at this time clearly going to products which were not being formed at an appreciable rate prior to the complete conversion of the NO. These most unusual changes in the NO_2 concentration also add challenge to the problem of mathematical simulation. They are also of great practical importance since several of the most damaging reaction products are not formed until the conversion of NO to NO_2 is complete. The author's own summary of the scheme necessary to account for these observations is shown in the attached figure. The heavy lines on the left represent the fast reactions 1, 2, 3 in Seinfeld's paper. These lead to the steady state relationships for oxygen atoms and ozone first clearly stated by Leighton and used here by Seinfeld. The other two thirds of the diagram show the

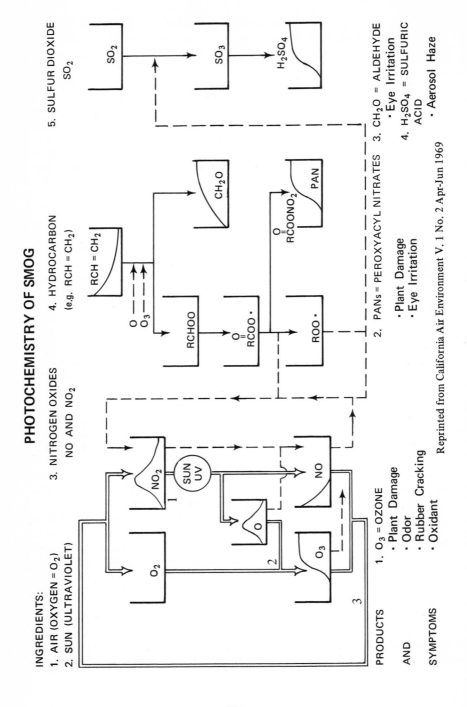

PHOTOCHEMISTRY OF SMOG

Reprinted from California Air Environment V. 1 No. 2 Apr-Jun 1969

204

oxidation of hydrocarbons as initiated by oxygen atoms and ozone. This scheme shows both formation of symptom-producing products and the oxidation of NO to NO_2 by peroxy radicals as well. Full computer simulation of this scheme is out of the question at least if the hundreds of different hydrocarbons simultaneously present were included. Many important rate constants can only be guessed at. The problem then is to choose a simpler set of reactions with suitable rate constants which will simulate the most important features of the atmospheric system. Seinfeld has chosen a very simplified scheme involving only generalized peroxy free radicals oxidizing the nitric oxide. His reaction 6 has been postulated for at least fourteen years in one form or another. The presence of free radicals, in fact peroxyacyl free radicals, is clearly shown by the formation of peroxyacyl nitrates,

$$\overset{\displaystyle O}{\underset{\displaystyle RCOONO_2}{\|}}$$

which must be formed by a chain termination reaction (as in Seinfeld's reaction 7) between peroxyacyl radical and nitrogen dioxide. But this is the only clearly identified product of a termination reaction. Alkyl nitrates, $RONO_2$, have frequently been identified in laboratory simulations of polluted air and in such systems their formation must be a significant termination reaction. But they become unimportant at truly realistic concentrations and were not detected even at 1 ppb in polluted air. Radical producing reactions are even harder to specify. They cannot be as simple as Seinfeld's reaction 4 because these reactants each have an even number of electrons whereas a free radical has, by definition, an odd number. Combination of an olefin molecule with either an oxygen atom (reaction 4) or an ozone molecule (reaction 5) must produce either two free radicals or none. To produce two free radicals requires the uncoupling of an electron pair and this generally requires energy. Of course the oxygen atom is initially in a triplet state with uncoupled electrons and this may facilitate the process. This is all speculation however; it is not even certain at O atom and ozone attack are the sole methods of olefin reaction. Furthermore, the olefins make up a small portion of the total hydrocarbon and although the paraffins and aromatics are less reactive they are present in larger quantities and surely make a contribution to total reactivity.

Seinfled's steady state equations for oxygen atoms and ozone are the same as derived by Leighton (1961). The argument for neglecting reactions 4 and 5 in their steady state is quite strong and not just an "assumption." Rate constants for these reactions are known and are much too small for them to compete with reactions 2 and 3. In fact the reaction of O atoms with olefin (reaction 4) would have to proceed much faster than the collision rate to compete with 2.

The steady state equation for free radicals is derived on the assumption that reaction 6 is a branching reaction which produces ϵ free radicals for each one consumed. As written, though, $\epsilon = 1$ in reaction 6 and there is no branching. As stated above, reaction 4 either produces two radicals or none. The steady state equation for free radicals, while it may be useful as a mathematical approximation, is not rigorously derivable from this mechanism.

Any model of the air photochemistry must include reactions 1, 2, 3, the three fast reactions which create the cycle on the left side of the figure. Any scheme which converts NO to NO_2 as an adjunct to hydrocarbon oxidation will account qualitatively for the observed changes in NO, NO_2 and O_3 *if* more than one NO is oxidized for each O atom and ozone molecule used to initiate hydrocarbon oxidation. Making a quantitative fit to real atmospheric data and inclusion of other active products in the scheme will require considerable study.

QUALITY STANDARDS FOR THE CONTROL
OF AIR POLLUTION

By Daniel R. Mandelker

Professor of Law

Washington University

St. Louis, Missouri 63130

Air pollution abatement presents yet another example of the American approach to environmental control. We isolate an evil, respond with a remedy which contains a fair amount of overkill, and are then surprised when results do not always match expectations. Like the control of housing quality through comparable standards, air pollution regulation intervenes in an area of economic activity in which the mix of costs and returns has been well-established over long periods. Any legal control which alters that mix by introducing a new cost or by subtracting a free good — and air pollution like housing undermaintenance is a free good — will naturally provoke intense reaction and resistance from those who are regulated. This fact of economic life lies at the bottom of any attempt to set air quality standards, but it is a fact of life which is more ignored than considered.

Because air pollution is viewed, in economic terms, as an externality producing harmful effects, it is subject to analysis by that body of economic thought which deals with externality theory and which usually appears in operational terms as an exercise in cost-benefit analysis. Now it has been said of economic theory that it purchases a high degree of elegance at the expense of a certain narrowness in point of view. And one of the variables excluded from and not considered in the economic approach to air pollution is the legal variable — how we translate the teachings of economic theory into a method of legal control. We lawyers, furthermore, have contributed to this gap, for we do very little work on the theoretic problems of modeling different types of legal controls, whether in a descriptive, predictive, or prescriptive sense. This paper makes a start at filling this omission by taking a brief look at the legal framework for air pollution abatement, and at those factors which influence its success or failure.

Some Constraints on Legal Intervention

Let us first examine some of the constraints on the selection of legal methods of control which arise out of the nature of the air pollution problem:

1. *Lumpiness.* Environmental quality has been said to be a continuum, not an absolute. We are not automatically healthy above one pollution level, and

sick below another. But air pollution control is limited when dealing with this health continuum because the correctives for air pollution are said to be lumpy. They require significant quantum jumps from one type of pollution corrective to another, so that working out standards along a continuum which permits a variety of adjustments to the control standard is not usually possible. The polluter either changes his fuel or he doesn't, for example. As a result, legal standards tend to over-regulate or under-regulate, and the creation of optimum states becomes difficult.

2. *Inverse relation of cost to benefit.* We are told that initial expenditures on pollution control will yield higher returns than later expenditures, that the ratio of cost of control to environmental improvement increases with the increasing success of control measures. This kind of incremental inverse relationship between cost and gain does not fit well with traditional legal methods of control, which generally assume a constant ratio between costs and benefits throughout the regulatory system. Housing codes, for example, do not distinguish between the comparative gains of putting in a central heating unit or of fixing the front steps. To do otherwise would require a sophistication in our legal structure which it has so far not managed to develop.

3. *Second-order consequences.* Any system of legal control which imposes a new cost will result in a shifting of that cost to some other sector of the economy unless the new cost is accompanied by a technology change that brings compensating benefits. Thus a ban on incinerators in New York City apartments simply led to a wave of voluntary shutdowns and the shifting of the refuse disposal problem to the public collection system. Unfortunately, the legal system is not very good at dealing with second-order consequences through primary regulation, with the result that second-order effects often cancel out primary gains. A good example are the problems created by the relocation of displacees in the urban renewal program.

4. *Weaknesses in the cost-benefit evaluation.* Students of cost-benefit approaches to problem-solving in the public investment arena are increasingly critical of the usefulness of this analytic tool as an indicator of policy solutions. In the field of air pollution, many of the benefits are non-quantifiable and often create corresponding dislocations — less air pollution means fewer laundries. So the entire exercise is suspect, especially when it can be biased through the intentional or unintentional exclusion of significant variables. This problem is made even more acute by weaknesses in data collection and information sources, and by difficulties in enforcement, a problem which leads us to consider the role of administration in the control system.

The Impact of Administration on Standard-Setting

Standard-setting and the creation of administrative enforcement systems are often considered as separate and discrete problems. What is not recognized is that the standards that are selected interact with the administrative system that is chosen, so that the character of each is dependent on the other. In the air pollution field, a set of quantitative standards with no variance opportunities will require field and perhaps court enforcement but they will not require an elaborate administrative apparatus. Conversely, qualitative, general standards will require extensive recourse either to administrative or judicial bodies for continuing interpretation. If a court were chosen, probably there would eventually be a cry in most large centers for a special "pollution court", which would not be too dissimilar from a quasi-judicial administrative agency exercising the same function.

We can approach the interaction of standards and administration from either side. From the standards side, what should be obvious, and what is implicit in what has been said above, is that a decision on the appropriate level of costs and benefits is implicit in the decision on standards. In other words, setting the pollution standard forces the polluter to accept a determined level of costs in return for a projected level of community benefit as output. Unfortunately, the difficulties of working with cost-benefit ratios in the pollution context, which have been mentioned above, make the setting of appropriate standards a tricky exercise.

From the administrative side, it would seem clear that we need a prescriptive model of an effective method of administration before we can decide on what standards to choose. As one example, lumpiness in the correctives available to abate pollution problems, together with a lack of confidence in our data, might suggest an administrative system in which remedies can be more effectively tailored to the needs and requirements of individual polluters.

In a recent provocative paper,* Professor Hagevik has elaborated a model of the administrative process as it might be adapted to pollution control, a model which points to a series of specific and detailed suggestions both for substantive standards and administrative techniques. Hagevik views the pollution control process as an exercise in bargaining between the control agency and the offending polluter, in which a binding decision is reached only after a series of incremental, negotiated steps. Standards are to be set by allocating to each polluter his share of airshed pollution damage, leaving to bargaining only the "facts" of the individual polluter's contribution. But factual judgment appears peculiarly suited to the quasi-judicial determination, a process which Hagevik labels as inconsistent with his bargaining approach. Moreover, I would guess that polluters would be much more anxious to bargain over the standard rather than over the application of that standard to individual cases. Yet standard-setting is a policy judgment which ought to be free of any bargaining taint.

Other types of air pollution legislation also fit the Hagevik model. For example, it is common practice to specify a pollution standard, and then to provide for a variance from that standard in cases of hardship. This is a

---

*Hagevik, Legislating for Air Quality Management: Reducing Theory to Practice, 33 Law & Contemp. Prob. 369 (1968).

regulatory device borrowed from zoning controls, where it has failed miserably. My guess is that the variance opportunity in pollution control legislation will simply provide a back-handed opportunity for the kind of bargaining which Hagevik envisages, but which may seriously undercut rather than reinforce the anti-pollution program.

My own intuition is that an incremental bargaining process is precisely what is not needed in pollution control, and experience in a related field, with the control of housing quality, bears out my evaluation. There, as in anti-pollution regulation, the legislative aim is often implemented by requiring capital improvements to the dwelling structure. Whatever the formal legislative structure in local housing codes, the dominant administrative mode has been negotiated enforcement, and almost everywhere in urban America the results have been pathetically limited. What Hagevik overlooks is that the decision-maker in the public regulatory agency, and the decision-maker in the private, polluting firm, are working from a very different conception of the decision-making process. The private, polluting firm, like the owner of substandard housing, is forced to arrange his mix of costs, prices, and output to show an overall profit. Else he will fail in business. Any intrusion on his private decision about cost levels will be fiercely resisted. The public administrator, on the other hand, works from a much more ambiguous success standard, for there is no equivalent measure in the public world which carries the same powerful force as the need to show a profit. Some observers of the administrative process would indeed go so far as to argue that there is no objective function in public regulation, and that survival of the regulatory program, even at the expense of stated legislative aims, is the paramount aim. When we add to this analysis the ever-present influence of political pressures, which have no counterpart on the private side, we are led to conclude that compromise rather than unbending commitment will be the accommodating adjustment of the public regulator.

What all this adds up to is that protracted negotiation will tend to favor the regulated rather than the regulator. Bargaining and negotiation only weakens the public side over time. That this has been the result in housing code enforcement is all too obvious even from the limited studies that have been attempted.

Some Suggestions for Regulatory Standards

What I have said so far has certainly not been encouraging. But the analysis does suggest some approaches to legal standards for air pollution control which may improve the performance of abatement programs. And the starting point, I believe, is the tentative nature of our knowledge, both about the impact of air pollution control measures, and about the types of technological changes necessary to bring them about. Regulatory programs in America appear to be endowed with a kind of Teutonic absoluteness that requires all offenders to be brought up to the highest possible program standards post-haste. Certainly housing codes have this characteristic, imposing as they do a high-level standard of housing quality which is to be uniformly applied throughout the community, and with all possible speed.

Given the economic constraints, this kind of program objective is impossible to realize, and given the tentative nature of remedial technology in the air pollution field, to aim in this direction would be most unwise a course of action. The problem is only complicated in air pollution by the wide variety of polluters, from the family car to the family incinerator. More tentative, partial solutions would seem preferable. We have already taken this tack in the automobile field, where new exhaust systems remove some, but not all, of the polluting by-products from engine combustion. It is a useful intermediate step to take while we continue to grapple with the problems of a more advanced engine technology.

To come more specifically to the regulations to be applied to fixed and on-site polluters, I am very much opposed to the "zoning" approach described above, which entails the setting of a fixed standard subject to a variance. This mode of regulation only encourages the worst kind of arbitrary decision-making in individual cases. My own preference is for another legal model which has proved quite successful in dealing with complex problems with an economic base, and that is the body of law known as anti-trust regulation. Here we have solved a complex problem of control by utilizing generalized standards subject to interpretation, both by a high-powered federal administrative agency and by the courts. Administrative and judicial solutions are tailored for individual companies and industries under the more generalized legal rule. It is a method of legal control with respectable legal precedent, and which fits well the uncertainties and variable enforcement settings that characterize air pollution abatement. This approach permits bargaining, if by bargaining we mean the opportunity to adapt generalized rules to particular circumstances. But it also requires definitive decision-making which will give meaning to the generalized standard over time.

Generalized standards and high-powered administration also require money, expertise, and a meaningful legal jurisdiction, which is why I also believe that the enactment and enforcement of satisfactory and workable air pollution standards will require increasing federal intervention and industry-wide, rather than community-wide or even state-wide, solutions. Again, the automobile case is telling. In the housing field, it is significant that the only housing rehabilitation that works is carried out under federal standards, subject to federal supervision, and with federal money. We not only need more federal direction and more federal involvement in air pollution, but more flexible and tentative methods for the control of our physical environment which will keep our options open as our knowledge and our technology improve. What appears to be the blessing of irreversible commitment often lives to become a curse.

Moreover, for a variety of reasons too complicated to fully articulate here, I believe that federal administration and court adjudication will strengthen rather than weaken the public hand. Court proceedings require a combative ritual which evens the sides, and which places a higher premium on commitment to objective than does the less visible regulatory process which is open to pressures, overt and otherwise. If industry-wide uniformity is an important key to the success of regulation, then industry-wide regulation at the federal level should overcome resistance resulting from competitive disadvantage arising out of local variations in standards and enforcement. Furthermore, the greater visibility of federal administrative process, the higher status of federal administrative agencies, and their ability to hold out greater career rewards to the successful public servant, should help to stiffen the back of the public regulator who faces a private rival spurred on by internal cost pressures.

Hagevik, an urban planner, states at one point that "(T)he most pessimistic conclusions one might draw is that the law and the legal system are in many respects incompatible with the scientific pursuit of optimal conditions under constraints of uncertainty."* I would dissent. But I would urge at the same time that experimentation, empiric research in enforcement problems, and above all more speculative thinking about methods of legal regulation are essential if our efforts at air pollution control are to bring satisfactory results.

---

*Id. at 398.

Comments on Mandelker, Quality Standards For
The Control of Air Pollution

By James E. Krier
Acting Professor of Law, UCLA

The lawyers' contribution to this Symposium was originally to be limited to the "Legal Basis and Methods of Standard Setting." As Professor Mandelker's paper indicates, however, law trained men are no longer content simply to prepare opinion letters on questions of legality; they want now to assume their share of the responsibility for ordering productive human behavior. It is primarily for this reason, I believe, that Professor Mandelker has sidled away from a narrow assignment to consider, as he does, the question of *administration* of standards, of *implementation* of standards — in short, of human behavior. This is not some grand cooptation by the lawyer of the behavioralist's world, but something far more practical: the recognition that standard-setting and standard-administration are intimately related concerns. This point recurs throughout Professor Mandelker's paper.

The focus of the Mandelker paper, I suspect, reflects something else about the "Legal Basis and Methods of Standard Setting" — namely, the fact that in some senses there simply isn't that much to talk about. The range of available methods and techniques (legal tools) is fairly well known and the legal basis for those tools has been quite clearly mapped. The frontier, on the other hand, surrounds the questions of what mixture of techniques, and what forms of administration, best *implement* our standards.

Let us be clear what we are talking about when we refer to air quality standard. This entire symposium is concerned with the translation of air quality criteria into standards. At least since the passage in 1967 of the Federal Air Quality Act,[1] those terms — criteria and standards — have quite precise meanings:

> "Air quality *criteria are descriptive* — that is, they describe the effects that can be expected to occur whenever and wherever the ambient air level of a pollutant reaches or exceeds a specific figure for a specific time period. Air quality *standards are prescriptive* — they prescribe pollutant levels that cannot legally be exceeded during a specific time period in a specific geographic area."[2]

As should be quite clear from this definition, little has been accomplished by the setting of air quality standards until some effective program has been established to implement them. The Air Quality Act itself recognizes this

---

[1]Pub. L. No. 90-148; 81 Stat. 485 (1967). The Act is codified in 42 U.S.C. §§ 1857-1857*l* (Supp. 1968) and further citation will be to the codification only.

[2]Hearings on S. 780 Before the Subcomm. on Air and Water Pollution of the Senate Comm. on Public Works, 90th Cong., 1st Sess. pt. 3, at 1154 (1967).

two step process, for it speaks not simply of state adoption of "ambient air quality standards" but of "a plan for the implementation, maintenance, and enforcement of such standards" as well.[3] And, as I have said above and as others have said before,[4] it is as to implementation and enforcement that creativity is especially demanded.

With these preliminary thoughts in mind, Part I of these comments will attempt to document briefly my claim that the legal basis of and present techniques for standard setting are well established by sketching the relevant "legal" questions and methods, with no attempt to be exhaustive; Part II will focus on Professor Mandelker's ideas and suggestions.

I
Legal Basis and Methods for Standard Setting

Legal Basis. When we talk about a legal basis for setting (and presumably implementing) some standard, we are concerned essentially with two concepts — one dealing with the general authority to act (by setting up and implementing, for example) and the other dealing with external constraints limiting the scope of that authority (the requirement, for example, that regulations be reasonable and not arbitrary). There is a relatively large body of literature addressed directly to the application of these concepts to air pollution legislation,[5] and I will simply summarize here its high points, most of which are devoid of any controversy.

The fundamental source of authority for state and local government to legislate concerning air pollution problems is the police power, a phrase incapable of precise definition but used and given shape and meaning by the courts in the course of thousands upon thousands of adjudications. The term identifies valid governmental actions undertaken to protect the health, safety and welfare of the community.[6] There is no question that air pollution control measures fall within the police power.

The legal (as opposed to practical, political, and so forth) constraints which limit the authority of government to act and thus make the police power something less than absolute are primarily the constitutional origin — the familiar concepts of due process and equal protection of the laws. Due process requires both "reasonable" and "certain" legislation, but these are

---

[3] 42 U.S.C. § 1857d (c)(1) (Supp. 1968). See also Stern, *Air Pollution Standards,* in 3 Air Pollution 601, 610 (Stern ed., 2nd ed. 1968).

[4] See, *e.g.,* Hagevik, *Legislating for Air Quality Management: Reducing Theory to Practice,* 33 Law & Contemp. Prob. 369 (1968).

[5] One of the better articles is Pollack, *Legal Boundaries of Air Pollution Control — State and Local Legislative Purpose and Techniques,* 33 Law & Contemp. Prob. 331 (1968). Much of the discussion of legal basis and method that follows is drawn from this article.

[6] See Sax, *Takings and the Police Power,* 74 Yale L.J. 1, n. 6 (1964). It is unusual to speak of a federal police power. The working equivalent for the federal government is the grant of power in Article III Section 8 of the Constitution.

broad and dynamic terms. Reasonableness requires only that the legislative body not act arbitrarily nor more broadly than necessary to accomplish its purpose. Since the courts recognize the right of the legislature to exercise wide judgment both as to what the facts of a problem are and what action might best be taken to meet it, this is a loose-waisted constraint indeed. So too with the requirement of "certainty." The utmost mathematical precision is by no means required; the essential question is whether the meaning of the statute can be reasonably ascertained, whether persons within the scope of the statutory requirements have been put on reasonable notice of the actions required of them.[7] Moreover, the proscription against vagueness has little impact outside the realm of criminal prosecutions.[8]

The equal protection doctrine precludes arbitrary descrimination but not reasonable classification based on the purpose of the legislation. Necessarily, the legislature has considerable discretion in setting up classes for differing treatment, especially in a complex field like air pollution control, where diversified sources and types of pollutants require varying forms of action. In short, neither the equal protection clause nor the due process clause unduly restricts the enactment of effective air quality standards. This becomes especially clear when we remember the judiciary's sensitivity to the legislature's need for broad powers in the regulation, for the public welfare, of essentially economic activities (as compared, for example, to legislation regulating more "personal" or "political" activities, such as voting, education or, indeed, procreation). That sensitivity is implemented, particularly in the economic area, by the doctrine that legislative enactments are presumed, in the absence of strong evidence, to be constitutional.

Legislative Methods. How have legislative bodies exercised their power to control air pollution? What methods have they used for standard setting and, more important, for implementation of those standards?[9]

Most commonly, air pollution standards have been concerned with emission controls (performance standards); Pollack points out that the bulk of

---

[7]See, *e.g.,* 50 Am. Jur. Statutes §§ 472-73. For an application of these principles in the context of air pollution control legislation, see *People* v. *Plywood Mfrs. of Cal.,* 137 C.A. 2d Supp. 859, 291 P.2d 587 (1955).

[8]See *Boutiliei* v. *Immigration Service,* 387 U.S. 118, 123 (1967) (void for vagueness doctrine is applicable to civil actions where "the exaction of obedience to a rule or standard . . . was so vague and indefinite as really to be no rule or standard at all . . .") There is, however, a developing doctrine that statutes can also be "void for vagueness" if they do not set forth sufficiently precise standards to guide administrative exercises of discretion. The doctrine's purpose is to preclude arbitrary official behavior.

[9]This brief suvey of method is drawn from Pollack, *supra* note 5 at 342-56. See also Edelman, *Air Pollution Control Legislation,* in 3 Air Pollution 553 (Stern ed., 2d ed. 1968).

The standards with which I am concerned here are essentially emission and specification standards, as opposed to ambient air standards. The latter are essentially declaratory, not proscriptive. See O'Fallon, *Deficiencies in the Air Quality Act of 1967,* 33 Law & Contem. Prob. 275, 278 (1968).

such legislation has been aimed at the density and opacity of smoke, although modern legislation commonly contains further limitations on particular pollutants, such as sulfur dioxide or particulate matter. Emission controls have a sound legal basis and incorporate the advantage (many think) of leaving to the owner of a pollution source the decision how best to comply with the law. Some jurisdictions, however, have gone beyond the setting of emission standards to regulate fuels, equipment, and even equipment operators (by requiring instruction in air pollution control techniques, for instance). And, of course, it is common for the legislative body to outlaw some activities altogether as a method of control; prohibition of open burning is an example.

Methods of implementation (or administration) vary. Permit and operating certificate systems have been widely used to implement equipment regulation. Such systems will almost uniformly rely on administrative enforcement in the first instance, with final enforcement resting in either civil or criminal court actions. Pollack reports that, as to enforcement of air pollution control standards generally, criminal actions are most common. There has been experience with civil injunction proceedings as a primary tool for enforcement, however, and opinions vary sharply as to which is the better technique.[10] Both are probably necessary.

This is not by any means a complete survey of legislative methods of standard setting. Nor, indeed, is "standard setting" (as we have used the phrase here) an exclusive method of air pollution control. Zoning, for example, could be used as an exclusive means of control, or in conjunction with performance and specification standards. In essence, zoning puts air pollution "downwind"; obviously, it is not the complete answer to air pollution, even as to that caused by stationary sources. It has been suggested that to be effective, zoning constraints "should be couched in terms of emission royalties or license fees graduated by location rather than by black-and-white prohibitions here and permissions there" and should be combined with a program of property taxes designed to encourage optimum use of downwind and upwind land.[11] Other methods of legislative control, of course, include a whole range of economic incentives and disincentives.

II
Some Comments

If there is little in the way of "legal" constraints on air pollution strategies, and if there is an abundance of method and technique, how do we

---

[10] See Walker, *Enforcement of Performance Requirements with Injuctive Procedure,* 10 Ariz. L. Rev. 81 (1968); Mix, *The Misdemeanor Approach to Pollution Control, id.* at 90.

[11] American Assoc. for Advancement of Conservation 295 (1965). On zoning generally, see *e.g.,* McGrath, *Planning and Zoning — Can They be Made to Work for Clean Air?,* Proceedings of Third National Conference on Air Pollution 554 (1966); Holland *et al., Industrial Zoning as a Means of Controlling Area Source Air Pollution,* 10 J. Air Pollution Control Ass's 147 (April 1960). These articles discuss land zoning. "Air zoning" has also been suggested: "allowing specified types and amounts of pollution to be emitted into a given body of air upon specified conditions. . ." See Air Conservation, *supra,* at 10, 56-57.

choose our package of standards? Professor Mandelker suggests that in attempting to answer that question we must remember that legal standards will be subject to "practical" (extralegal) constraints that must be kept in mind in choosing a legislative program, and that they also will gain or lose effectiveness in a real world depending upon the administrative machinery set up to implement them. It would be difficult to disagree with these suggestions, and I would like simply to add to them and, perhaps, to question what they say about Professor Mandelker's conclusions.

Professor Mandelker's list of constraints on legal intervention does not pretend to be exhaustive, and I would suggest that what can be called a socio-political or public-opinion constraint must be added. This is a complex, complicated limitation on effective law making, and I can only suggest what I see as a few of its facets. First, there is no question that most of us have an apparent interest in the continuation of air pollution, because we wish to drive a car (or two cars) or burn rubbish and leaves or pay the lowest prices for products or make the highest profits in our businesses. This does not mean that individuals do not also have a real interest in ending or diminishing pollution, but it does mean that this is not an exclusive or all-embracing interest; and it generally means that voters will support programs aimed at *them*, and that they will support programs which *appear* to involve the least cost for them. Surely this is one (although not an exclusive) reason for legislative and popular focus on industrial emitters. As voters learn, however, that such a focus means higher costs to them in the long run, and as they learn that other measures imposing more direct costs on them will be necessary, popular support for further legislative action may be sorely lacking.

This suggests a need for broad public educational programs designed to convince the public that the costs they must bear are outweighed by the benefits of clean air. But, we learn on every front, we have not yet reached a convincing state of precision in regard to cost-benefit appraisals. Professor Mandelker's paper gives some evidence, I think, of our lack of sophistication here. He makes reference in his paper to the common argument "that initial expenditures on pollution control will yield higher returns than later expenditures, that the ratio of cost of control to environmental improvement increases with the increasing success of control measures." Yet how can he or anyone else make this statement with confidence and at the same time point out, as he does, the many quantitative and qualitative weaknesses of cost-benefit analysis? For example, a common expression of Mandelker's first point is that it may cost X dollars to cut emissions from a particular source by ninety percent, and another X dollars to cut them by five percent more. But the error in arguing that this produces a declining level of benefits occurs in equating benefits to the environment with percentage decreases in pollutants. What if the additional five percent decrease meant that we avoided crossing some threshold of biological tolerance beyond which we could not live (as an extreme example) or what if it meant we gained some sense of aesthetic enjoyment which could only be achieved by removing the very last vestige of an offensive phenomenon? Could we then say that there was nevertheless an inverse relation between cost and benefit? This is simply to underscore Professor Mandelker's point that cost-benefit analysis may in many instances not get us far. We may lack the tools for making the real measurements upon which helpful analysis depends.

What often happens in such situations is the replacement of facts with sermons and panic-mongering. It is common today, for example, to hear that we simply have no more time, to hear of outrageous environmental abuses and insults by large industry, to hear that we must have "clean air at any cost." Yet common sense tells us that this simply is not so, that there is time to take reasoned, well-informed action. And, at least subconsciously, our common interest in avoiding costs reinforces this common sense belief. "Things seem to be going along well enough; I can drive my car for another year or so." We engage in incremental avoidance of costs, and (because we do not yet have the capabilities for proving otherwise) we *seem* to get away with it. We keep driving cars, and we keep breathing — and living.

In short, the sort of wolf-crying in which many well-meaning reformers engage creates tremendous credibility gaps which often make passage of rational legislation extremely difficult, especially if it imposes direct costs on voters. Sermons have another danger; like most arguments based on moral outrage, they tend to be myopic. We attack dirty air with a zealousness that overlooks the fact that many proposed solutions would only produce pollution of the water, or the land. (A ban on the use of incinerators for the disposal of garbage, for example, might reduce air pollution but result in a random dumping of waste in rivers or open fields—or in the streets.) We concentrate all of our efforts on the killing effects of air pollution — on damage to health and life. In doing so, we understate and ignore and thus fail to provide support and precedent for aesthetic considerations. Our escape from this is to continue techniques of fact gathering, analysis and dissemination, but at the same time learn to focus more attention on nonquantitative concerns (such as aesthetic considerations) in the context of a whole environment. Certainly we cannot stop the rumor-mongering and, in any event, it probably has some value in counteracting the equally zealous and polar position of many industrial interests which fight constantly to understate the pollution problem.

This leaves open, of course, the question of what we do with the information we have. How do we translate it into effective action (and law)? As Professor Mandelker points out in his paper, an interesting answer to this question has been modeled by Professor Hagevik in a recent article.[12] Professor Mandelker has briefly outlined Hagevik's proposal; in essence, the proposal suggests that bargaining theory would be productive as a psychic supplement to the minimum-cost concerns of the welfare economists. Hagevik senses a national commitment to direct regulation rather than to effluent fees or direct subsidies, either of which may be more ideal theoretical measures for

---

[12] Hagevik, *Legislating for Air Quality Management: Reducing Theory to Practice,* 33 Law & Contemp. Prob. 369 (1968).

dealing with air pollution diseconomies, and has shaped his proposal accordingly. He points out, however, that a program of direct regulation does not foreclose some reliance on "subsidies, licenses, permits, effluent charges, emission standards and variances therefrom, emergency powers, and some reliance on market forces."[13] One of the main objectives of Hagevik's proposal is to provide for incremental decision making and experimentation. (It may be an interesting irony that Hagevik promotes incrementalism, yet proposes a legal model so different from existing legislative methods as to be much more than an incremental next step for pollution control. That is, is Hagevik thinking incrementally?)

Professor Mandelker claims to differ with Hagevik, but I am unsure that he does in the final analysis. While he rejects "an incremental bargaining process," he also suggests avoidance of "Teutonic absoluteness," praises the "useful intermediate step" of some automobile pollution legislation, and suggests a program (modeled on anti-trust regulation) which would utilize "generalized standards subject to interpretation" with final solutions "tailored for individual companies and industries under under the more generalized legal rule." And Professor Mandelker admits (as it appears he must), that "this approach permits bargaining." Thus the only real point of divergence between him and Hagevik appears to be whether (and what kind of) general rules will be the ultimate authority against the background of which bargaining would occur.

Much of Professor Mandelker's skepticism about present approaches to air pollution control and about the Hagevik model grows from the American experience with housing code enforcement. But is that experience relevant here? The premise (at least implicit and often explicit) of most housing code legislation has been that either landlords or tenants could afford to pay the costs of standard housing. Yet, we are beginning to learn today that, at least in the case of low-income housing (where the problem is most acute) this simply is not so: many landlords are operating at the minimum profit required by the risks of the business and, further, cannot pass the costs of required improvements on to tenants, who already are paying more than they can afford for housing. Thus, we are learning, massive public subsidies are called for. Are we faced with the same problem in respect to air pollution? Can't the increased costs of control devices be passed on to consumers to solve at least part of the problem? Moreover, aren't some of the costs of air pollution control techniques and capital outlays actually profit producing in that they produce economically valuable by-products which can themselves be marketed? (Ordinarly the landlord of low-income housing cannot realize on his capital improvements, for the low-income houseing resale market is very weak.)

What these questions suggest to me is that we must be quite selective in surveying the relevance of past experience. How much of the air pollution problem is like other problems with which law has dealt, and how and why were the legal measures designed for those problems effective and ineffective?

---

[13]*Id.* at 379.

How much of the air pollution problem is unique? Is air pollution like other "environmental" problems? Some success with effluent charges as a solution to water pollution in Germany, for example, has resulted in advocacy of effluent charges for air pollution. But are the problems the same? It has been suggested that they are not,[14] so different measures may be called for. Professor Mandelker suggests something after the model of anti-trust regulation. But how successful has anti-trust regulation been? Can we possibly measure the impact it has had on reaching the ideal of free competition (putting aside the question of whether we want to reach that ideal); how do we know, for example, the percentage of secret conspiracies we have ended; what do we know about the effect of the legislation and its sanctions on business behavior?[15]

These comments are by no means meant to be negative. Professor Mandelker has pointed out that we are working in an area about which we know little, and it is important to remember that the shortcomings in our knowledge relate not only to technical matters (about the effects of certain pollutants, the cost of various abatement procedures, and so forth) but to matters of human behavior as well. This may affect our strategy for the next decade. I am prone to agree with Professor Mandelker that increasing federal intervention is ultimately necessary to manage our air pollution problem most effectively. But I would hope that that intervention would not be at the price of losing the practical knowledge we might gain from having fifty different programs of enforcement, rather than one uniform one. In short, it may not yet be the time for further federal intervention. In the meantime, flexible methods of approach — whether after the Mandelker model, the Hagevik model, or some other — are certainly called for. The important thing is to experiment and learn while we can.

---

[14]See Kneese, *Air Pollution — General Background and Some Economic Aspects*, in The Economics of Air Pollution 23, 33 (Wolozin ed. 1966).

[15]See generally Ball & Friedmann, *The Use of Criminal Sanctions in the Enforcement of Economic Legislation: A Sociological View*, 17 Stan. L. Rev. 197 (1965).